ETERSEN'S BOOK OF MAN IN SPACE

A NEW ENVIRONMENT

olume Two

MAN IN SPACE Vol. 2 A NEW ENVIRONMENT

Edited by Al Hall and the General Subjects editors of Specialty Publications Division.

ACKNOWLEDGEMENTS

The editors of Petersen's MAN IN SPACE are indebted to the following agencies, companies, and individuals for their enthusiastic cooperation in helping locate and make available the numerous photos and hundreds of bits of information which add immeasurably to the scope and content of this volume. Our profuse thanks to: the National Aeronautics and Space Administration (which supplied all photos unless otherwise credited), and especially Les Gaver, Margaret Ware and the Audio-Visual Branch, NASA Headquarters; Jack Stewart, Karla Scott, Donna Green, and NASA's Ames Research Center; John MacLeish, Jack Riley, Jim McBarron, and NASA's Lyndon B. Johnson Space Center; Barbara Rogers, Dorothy Marine, and NASA's John F. Kennedy Space Center; Edgar Russell, Jr. and IBM's Federal Systems Division; Raymond A. Deffry, C.W. Birnbaum, and McDonnell Douglas; James D. Grafton and Boeing Aerospace Company; P.G. Smith and Aero Spacelines; Frank Colella, Frank Bristow, Bob McMillan and the Jet Propulsion Laboratory of Cal Tech; Harry Lowenkron, Laura Melin, and Sovfoto; Don Bane and the Public Relations Department of TRW Space Systems Group; Mike Gentry and Technicolor, Inc.

COVER

First U.S. space walk. Astronaut Edward H. White II is photographed by astronaut James A. McDivitt during the historic Gemini 4 mission, June 3-7, 1965. NASA photo. Cover design by Pat Taketa.

INSIDE FRONT COVER

Nations around the world commemorate the efforts of the United States and the Soviet Union in space exploration with postage stamps (From the collection of Kurt Raymond Hall)

Library of Congress Catalog Card Number 74-82253 ISBN 0-8227-0073-5

CONTENTS

INTRODUCTION

Welcome to the second in this five-volume series of MAN IN SPACE. Volume Two chronicles the flights of Project Gemini, or roughly the era from late 1963 through late 1966. It was a period of great agitation not only for the space program, but for the nation as well. These were the critical years, too, critical in terms of the number and variety of the problems which had to be solved both in space and on earth.

THE KEY IS COMPETITION

The splashdown of L. Gordon Cooper's *Faith 7* capsule brought the manned mission phase of Project Mercury to a successful conclusion. America had proven, that like the Soviet Union, it could also send a man into orbit and recover him successfully. But between the conclusion of Cooper's flight and the first manned Gemini launch, which would introduce America's two-man spacecraft and develop technology necessary for a manned lunar landing, came reports that Russia was already testing a three-man spaceship with a new and very powerful launch booster. Were the Russians aiming at the moon already?

Those who weren't concerned with where the Russians went were classified as isolationist holdovers from America's not-so-distant past; the same holdovers who also thought that the United States could keep to itself and ignore that itinerant paperhanger who called himself Adolf Hitler. America could not keep to itself, of course, because willing or not it had become a world power. It had reestablished a form of government called "a democracy," the motive force of which was free enterprise known as capitalism. It was the technological advances made under this form of government that made the world sit up and take notice of America and eventually pushed the country into its position as a world leader.

The Soviet Union emerged from World War 2 as far more powerful than when it had entered. But its form of government, barely 28 years old at that time, was the opposite of America's. It was a type of government in which free enterprise was replaced by collective endeavor with all proceeds to the State and all benefits coming from the State. This system was called communism.

Naturally there are other forms of government, but these two, being most powerful, were also very popular. Thus, it was inevitable that one power would attempt to dominate the other. Each tried to gain an advantage in quantity and quality of weapons, vastness and complexity of technology, and so forth, but neither could achieve any clear-cut superiority. A strange kind of war thus developed.

ENTER THE COLD WAR

It was called a Cold War. What is a Cold War? Simply stated, a tug-of-war between two ideologies: America's capitalism and Russia's communism—with each side trying to attract interested spectators (uncommitted nations) to join in and pull for their side. Showing these nations of spectators that your side offers "more" is perhaps the greatest inducement for them to adopt your form of government and join you. That is why Sputnik had such a profound impact. It showed the world that communist Russia's technology was equal-to or superior-than democratic America's. It also had a tremendous impact on America, because it took on the aspect of being the clear-cut advantage that might precipitate another world war. Just the thought of a Soviet base in orbit or (worse yet) on the moon was enough to send a chill through every military strategist in Washington, D.C.

President Dwight Eisenhower's administration had moved to meet Russia's challenge of military arms and missiles by massive defense spending, but showed only moderate interest in meeting Russia's space challenge. In fact, Eisenhower's chief aide, Sherman Adams, stated that "...this administration is not interested in a high score in an outer-space basketball game." It was a rather strange attitude for an administration that labored so hard to make sure a Russian airplane was matched in size and power by an American airplane, or a Russian Intercontinental Ballistics Missile was matched in range and destructiveness by an American ICBM!

JFK ACCEPTS THE CHALLENGE

When John F. Kennedy became President, he replaced Eisenhower's "Fair Deal" with "The New Frontier." One of the new frontiers that Kennedy's administration would push back was, of course, space. On May 25, 1961, John F. Kennedy called for the nation to commit itself to space exploration by landing a man on the moon. On November 22, 1963, just months after the successful conclusion of Project Mercury with the splashdown of Cooper, Kennedy was struck down by an assassin's bullet in Dallas, Texas. Vice President Lyndon B. Johnson, a strong advocate of America's space program, picked up the reins of Kennedy's administration. The space program forged ahead, still aiming at the goal Kennedy had chosen.

LBJ CARRIES ON

President Johnson and his administration knew full well that the goal was far more than a manned lunar landing before the end of the 1960s. It was also designed to prove to the uncommitted world that the United States of America was not a second-rate power and that she had not taken a technological back-seat to the Soviet Union.

That, however, was going to take a considerable amount of proving, for the Russians were piling up an impressive string of "firsts" in space exploration. America was proceeding with caution, because space—that blackness beyond earth's protective atmosphere—is a new environment for man: a very unforgiving environment where one tiny mistake can be instantly fatal. There were no reasons that would lead one to believe the Russians were not also being just as cautious. The progression for both countries was a logical one: First, unmanned vehicles are sent to test spacecraft design; second, animal subjects verify the suitability for man; and finally, man, himself, must be thrust into the unknown.

And unknown it was! There was the full fury of the sun, unscreened by atmosphere, just beyond the protective skin of the spacecraft; and the Van Allen Radiation Belt, discovered by America's Explorer One satellite. There were countless other hazards that, perhaps, man couldn't even conceive. But the timid probings of Project Mercury proved that short-duration space flights produced no detrimental effects on man. Yet what of sustained periods in space that would be necessary if man were to attempt to reach the moon or beyond? Would solar radiation, over a longer period of time, leave him sterile? Or, worse yet, cause him to produce hideously deformed offspring? In the era under discussion here, no one knew the answers and one simply doesn't rush into this type of exploration until one has more data. Then, too, man must learn how to do more than perform simple maneuvers in a spacecraft. A trip to the moon or establishment of a space station in orbit above the earth requires that two craft meet and join up. It would mean that perhaps even going outside the safety of the spacecraft would be necessary. All these and more were the goals set for Project Gemini. But while America was creeping cautiously toward the first two-man launch and the answers to some of the questions just posed, Russia was announcing one space "first" after another. And many Americans were asking themselves, "How we could really be that far behind if Russia has to take the same path to space as us?"

BURN, BABY, BURN

Other Americans, though, were asking themselves and others why we were trying to solve problems in space when there were more than enough problems right here on earth that begged for immediate solutions. Equal rights, guaranteed by the Constitution of the United States, was a case in point. Any member of a racial minority could tell you that those words were hollow. Civil rights leaders urged America's Negroes to display their feelings about segregation; and protest marches were organized on scales never-before seen. Alabama and Mississippi saw many of these demonstrations, but they were by no means limited to the Deep South. Northern and Eastern states saw more than their share, too. Many demonstrations that began peacefully ended in full-scale riots. The mood of the nation was such that when Los Angeles police stopped a young Negro during a routine patrol in the south-central ghetto area known as Watts, it precipitated one of the worst riots in modern American history. A small crowd quickly became an angry mob; stones hurled at police turned into fire bombs that were thrown into stores. Isolated fires soon begat a raging holocaust that virtually consummed South-Central Los Angeles and burned for days, kept alive by cries of: "Burn, baby, burn!" And the heat of those fires in the summer of 1965 kindled others across the nation. But out of the turmoil grew a new pride; a new identity and the replacement of the word "Negro" with the word "Black," uttered proudly.

Yes, not only was the era from the end of Project Mercury through the end of Project Gemini critical for the space program, it was critical for the nation, too. The spectre of war in Vietnam was growing in cost, both in American lives and American dollars. Riots protesting America's presence in Vietnam took place in countries around the world and American students held protest rallies on college and university campuses across the United States.

Strikes and inflation also dulled the lustre of President Johnson's "Great Society" which he promised when winning the Presidential election of 1964 over Republican nominee, Barry Goldwater. The prosperity of the Eisenhower and Kennedy years was being consummed by spiraling costs which then triggered labor strikes for higher wages.

THE WORLD'S BURNING

The United States was not the only country beset by woes. During the era mentioned, the Dominican Republic suffered a violent change of government; as did Syria and Argentina. Most of the young nations of Africa suffered through some form of internal strife during this era, as did Greece. Brazil, too, had its governmental problems. And even Communist China was the scene of government purges. The Soviet Union? Yes, even they had problems...with the Communist Chinese!

This then, was the state of the nation and the world in the turbulent early-and mid-Sixties that were the setting for man's second round of explorations into space. Viewed in the perspective of man's history on earth, it was simply "the same old world." Man was still fighting, still bickering and still looking for something a bit better.

A CAUSE FOR SPACE

Could it possibly be that by exploring space man might find the answers he seeks? And if he does, will he be able to recall exactly what it was that prompted his journey? Will he be able to look back and say, "I remember well those early days of the space program!" This series might help to jog the memory, by recounting step-by-step those projects which started man down that different and difficult road.

In recounting the story of man's exploration of space it is difficult to actually isolate one space project from another, because its very much like a football game. The forward pass that results in a touchdown is merely the culmination of many individual efforts directed toward that moment and superseded by the very next drive for a touchdown.

Each space mission, manned or unmanned, is very much like that touchdown. It has its moment of glory posted on the scoreboard until the next mission increases the score and shines for all to see until it, too, is eclipsed by yet another mission that posts a higher score. Just about everyone interested in the game hears of the final score, but it's more fun to know how the score was compiled and against what odds.

OUR APPROACH

All this is by way of trying to explain in simple terms, a very complex operation that involves a "team" comprised of hundreds of private contractors and government agencies plus hundreds of thousands of employees. Liken launch and mission control personnel to the quarterback, the receiver to an astronaut (or astronauts) and the linemen to support people (government and civilian) and you'll begin to have a better concept of our space program. NASA would, naturally, be equated with the coaches and front office, because while the team is playing one game, they are monitoring it and planning ahead for the remaining games and, eventually, the championship.

It would be next to impossible to describe each and every person and company and list their contributions, so attention is focused on the "stars" of the show: The launch and mission control people and the astronauts. But just like the televised pro football games, the time between quarters, or even downs, can be utilized to point out the efforts of the linemen and coaches. This is basically how Petersen's *MAN IN SPACE* presents America's space program. We are not giving you the filmed highlights of a few of the best plays from one or two of the season games and then putting all our emphasis on the championship game. *MAN IN SPACE* is a moment-by-moment chronology of the "team's" efforts from the first Spring scrimmage through the post-season finale! And we'll shift our cameras and commentary from the action when its appropriate to take you behind the scenes for a look at the other, and often much neglected, members of the "team."

Oh yes, just like on television, we're not going to tip off the big plays and spoil some of the excitement for you fans who couldn't watch it live. We'll tell you about it as though it were happening at that very moment and let you watch the quarterback throw the bomb on "third-and-two" and see the receiver sprint under it with outstretched hands.

THE EDITORS

As our manned space program gathers momentum, the recruiting and training of new personnel for the astronaut ranks has become an important phase of NASA operations. Parameters set for the original seven men are slightly changed. They now may be six feet in height, but still must meet exacting physical standards set up by the Astronaut Office, particularly in regard to weight, eyesight, hearing and physical coordination. These factors, in conjunction with the professional requirements and training regimen, make the prospective astronaut in all probability a very seasoned pilot with experience in supersonic aircraft. It has been concluded that a second group of astronauts should consist of nine men, competitively selected after meeting the basic requirements. In September, 1962, NASA announced the selection finalists: the "new nine." Photographed together around a Project Gemini capsule model (**1**), they are from top left, reading clockwise: Neil Armstrong, a NASA test pilot and former Naval

THE NEW NINE

Introducing the latest groups of astronauts

1

aviator; Frank Borman, a USAF fighter and test pilot; John Young, a Navy fighter and test pilot; Thomas Stafford, an Annapoils graduate and USAF aerospace research pilot; Charles Conrad, Jr., a Naval aviator and test pilot; James McDivitt, an Air Force test pilot and combat aviator; James Lovell, a Navy test pilot; Elliot M. See, Jr., a former Naval aviator and civilian test pilot; Edward H. White II, a USAF fighter and test pilot. Aside from their status as pilots, all nine men share a high educational level, holding degrees from either our military academies or universities. All are specialists in either engineering or military science. All nine are married and have families. Ed White, left (**4**), and Jim McDivitt share a light moment with Lt. Dee O'Hara, an Air Force nurse who is connected with the astronauts' medical staff. One factor that is of the utmost importance for every astronaut is physical condition: it must be peak at all times. Tom Stafford (**3**) works out frequently at the Manned Spacecraft Center gym along with the other astronauts. Each man has a specific physical fitness program assigned to him. Stafford, passed over once by NASA due to his 72-inch stature, now trains rigorously among the second group of astronauts. Despite the continuous physical training program delineated by NASA's doctors, the astronauts are by nature an active and aggressive group of men, participating in a variety of sports of their own accord. In hard hats (**2**), the men line up with Cape personnel during one of their engineering orientation sessions.

4

THE NEW NINE

Future pathways into space will rely on hardware that is more accepted in a general sense than the spacecraft of today. Beyond our lunar voyages, space exploration will take on an even greater emphasis toward understanding the universe we live in. The scientist is the man who has dedicated his entire life to research into our universe, be it in a social, physical or natural sense. This philosophy is reflected in the fourth group of astronauts chosen by NASA, a highly select group of five men. They have a slightly revised title to their new profession: Scientist-Astronaut. Seated, from left to right (**1**): F. Curtis Michel, a physicist; Harrison H. Schmitt, astrogeologist; Joseph P. Kerwin, MD, physician. Back row, left to right, are Owen K. Garriot, physicist, and Edward G. Gibson, physicist. Each of these men have been selected on the basis of their academic and research accomplishments. Their selection, announced on June 29, 1965, will be followed by a training period that will include flight in jet aircraft. Michel, who is

1

4

5

from LaCrosse, Wisconsin, is a former USAF pilot. Schmitt, born in Santa Rita, New Mexico, received his PhD. from Harvard, and is another avid sports enthusiast, including skiing, hiking, fishing and handball in his spare time. He is a bachelor. Joe Kerwin hails from Oak Park, Illinois, and is a Captain in the U.S. Navy Medical Corps, where he earned his wings. Garriot, a former Naval officer, is from Enid, Oklahoma, and maintains FAA commercial pilot ratings. Gibson, from Buffalo, N.Y., came to NASA from Philco's Applied Research Laboratory. Being an astronaut means maintaining a constant level of high professional knowledge. Here (**2**) astronauts Lovell and Aldrin study technical material related to the Gemini capsule simulator in the background. Flight proficiency is vital, too. Though one operates within the atmosphere, and the other outside, the qualifications for either overlap. Charles Duke (**3**) a USAF pilot and astronaut, poses in front of his T-38 Talon before a flight. Duke, selected in April 1966, comes from Charlotte, N.C., and graduated from the U.S. Naval Academy. Ed White jogs the Cape coast (**4**), while Richard Gordon and Clifton Williams practice desert survival (**5**). This is the fifth astronaut group (**6**), selected April 4, 1966. Seated, from left: Edward Givens, Edgar Mitchell, Charles Duke, Don Lind, Fred Haise, Joe Engle, Vance Brand, John Bull, Bruce McCandless. Standing, from left: Jack Swigert, William Pogue, Ron Evans, Paul Weitz, Jim Irwin, Gerald Carr, Stuart Roosa, Al Worden, Tom Mattingly, Jack Lousma.

3

THE NEW NINE

Jim Irwin (**4**) was born in Pittsburgh and graduated from the Naval Academy. He also holds an MS from the University of Michigan in Engineering. Tennis is only one of his recreational activities. Astronauts, from left, Lovell, Stafford, Givens, Cernan and Scott during swimming instruction (**2**). They look as if they are about to douse the instructor! Dick Gordon (**3**) with his wife Barbara and their six children. Eleven scientist-astronauts are added to the ranks in July, 1967 (**1**). Seated, from left: Philip K. Chapman, Robert A. Parker, William E. Thornton and John A. Lewellyn. Standing, from left: Joseph P. Allen, Karl G. Henize, Anthony W. England, Donald L. Holmquist, Franklin S. Musgrave, William B. Lenoir and Brian T. O'Leary. Again, these men have been chosen on the basis of their scientific or medical qualifications in preference to their ratings as pilots. The educational level of this group is impressive. O'Leary, Parker and Henize are astronomers, Chapman holds a doctorate in Instrumentation, while Lewellyn, born a British

1

4

subject, is a chemist. Thornton is an MD, while Holmquist is a physiologist and Musgrave holds degrees in both biophysics and medicine. Lenoir holds a degree in electrical engineering. These are the seven men selected as the seventh group (**5**) of astronauts to train for the Manned Orbiting Laboratory in the early 1970's. From left: Karol J. Bobko, born in New York, is a Lt. Colonel in the USAF and a member of the first graduating class of the Air Force Academy; Gordon Fullerton, who is from Rochester, N.Y., is a major in the USAF., and holds degrees in mechanical engineering; Henry Hartsfield, from Birmingham, Alabama, is an Air Force officer with degrees in engineering and astronautics; Robert Crippen is a naval officer from Beaumont, Texas, and has extensive flight experience; Donald H. Peterson is an Air Force Lt. Colonel and holds a MS in Nuclear Engineering; Richard Truly, a Navy flight officer with a Bachelors degree in Aeronautical Engineering, is from Fayette, Mississippi; Robert Overmeyer, a Marine pilot and Lt. Colonel, comes from Lorain, Ohio, and holds a MS degree from the USN Postgraduate School in Aeronautics. The astronauts share many similarities in background and education, and an interesting fact is that most were involved with the Boy Scout program during their school days, a number reaching the Eagle Scout level. Each of these men represents a considerable investment in the future of America and is well equipped to aid in our quest for more knowledge about space.

3

VALENTINA TERESHKOVA

The first woman in orbit

On June 14, 1963, at 3:00 PM Moscow time, the Soviet Union launched cosmonaut Lt. Col. Valery F. Bykovsky into orbit aboard Vostok 5. This flight was not unexpected by western observers, since the Nikolayev-Popovich double Vostok flight of ten months previous had shown the Russians were continuing to expand their space capabilities. What was quite unexpected was the launch two days later, at 12:30 PM Moscow time on June 16, of Vostok 6, piloted by this remarkable woman, 26-year-old Valentina Vladimirovna Tereshkova (**1**). Valentina's moment of glory was a far cry from her childhood. Born on March 6, 1937, in the village of Maslennikovo near Yaroslavl, she and her brother and sister were raised in hardship by their mother, after their father died in action early in World War II. She was not able to start school until she was ten years old, and at seventeen was apprenticed to the Yaroslavl Tire Factory. She left a few months later to join her mother and sister at a textile mill, where she became a loom op-

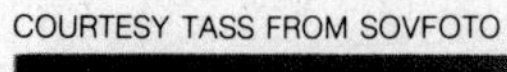
COURTESY TASS FROM SOVFOTO

1

erator. Her industrious nature soon asserted itself, and she became very active in two fields which greatly affected her future: Komsomol (the Young Communist League), and the unusual sport of parachuting. She pursued both activities avidly, and in 1960 was elected secretary of her Komsomol branch and received a first-class certificate as a parachutist after making more than 125 jumps. As was everyone else, Valentina was thrilled by Yuri Gagarin's flight in April of 1961, and she audaciously wrote to the Soviet Academy of Sciences and volunteered to be a cosmonaut, citing her ability in parachuting. To her surprise, she was accepted, and a whole new world opened. Her training began in March 1962. Gagarin, whom she at first tended to hero-worship a bit, was extremely impressed by her dedication, physical stamina, and long hours of study. Apprehension shows on the day of her flight (**2**), but once in orbit she becomes ecstatic. "I see the horizon," she reports during her first orbit, "a light blue, a blue band. How beautiful it is!" Her dual flight with Bykovsky continues for three days, the two Vostoks once approaching within three miles of each other on June 17. Shortly before 11:00 AM Moscow time on June 19, Valentina begins her re-entry. When into the lower atmosphere, she ejects from the capsule and parachutes down, landing safely at 11:20 AM. Her landing spot is in Kazakhstan, near a small village 380 miles northeast of Karaganda. Naturally all work stops as the excited villagers gather around their visitor (**3**).

COURTESY TASS FROM SOVFOTO

COURTESY TASS FROM SOVFOTO

TERESHKOVA

Valentina's 48-orbit flight has lasted 70 hours and 50 minutes, the longest completed space flight. Her record does not last long, however, for less than three hours later Bykovsky also lands safely about 330 miles northwest of Karaganda, having completed 82 orbits. The two are reunited the following day, and on June 22 a massive celebration is held in Moscow's Red Square (**1**) to honor them. Appearing a little like a rose among thorns, Valentina is the center of attention on this occasion (**2**), as she laughingly accepts the tributes of her fellow cosmonauts in the place of honor atop Lenin's tomb. The pilot of every Vostok is here: (from left) Popovich, Nikolayev, Titov, Valentina, Gagarin, and Bykovsky. Valentina is greeted warmly by Premier Khrushchev, who takes the opportunity to twit America for its bourgeois notion that woman is the weaker sex. Indeed, as he points out, her flight was longer than those of all the Project Mercury astronauts combined. On the same day, she is awarded her nation's high-

1

COURTESY TASS FROM S

est honor by being proclaimed a Hero of the Soviet Union, and from Leonid Brezhnev, Chairman of the Presidium, she and Bykovsky receive the Gold Star medal and the Order of Lenin, both traditionally awarded to returning cosmonauts. A few days later she holds her own press conference (**3**) in the House of Unions' Hall of Columns in Moscow for delegates of the World Women's Congress, at which she is joined again by her space partner, Bykovsky. In the west, there is speculation that the objective of these flights may have been a rendezvous. Valentina's flight is a bright success, and she handles her worldwide acclaim with a delightful modesty. Perhaps her most remarkable attribute is her ability to remain unmistakably feminine in a masculine operation. Though she holds the military rank of junior Lieutenant, in public she invariably wears a simple dress rather than a military uniform, preferring to look like a woman rather than a mass-produced officer. She uses little make-up, looking the better for it, and her tawny hair is rarely covered with a hat. She looks, in fact, exactly like the image of the ambitious, industrious, and naturally attractive woman that the Soviet Union likes to portray as the epitome of its young adults. On November 3, 1963, Valentina becomes Mrs. Andrian Nikolayev, and the two cosmonauts later become the proud parents of a daughter, Alyona (**4**), born on June 8, 1964. This child of two space travelers ends the speculation that human sterility may result from exposure to radiation in space.

COURTESY TASS FROM SOVFOTO

3

COURTESY TASS FROM SOVFOTO

4

COURTESY TASS FROM SOVFOTO

THE MANNED SPACECRAFT CENTER

A guided tour of NASA's facility in Houston

Situated on 1620 acres of Texas pastureland is NASA's base of operations for all manned flight operations—Manned Spacecraft Center. "MSC," as it is known, is a complex of administrative, operative and research buildings and facilities (**1**) whose function is the development and control center for our space program. Manned Spacecraft Center is located some 25 miles southeast of Houston, and despite the distance, has linked that city indelibly with the space program; "Houston Control, this is....." has become a familiar call to fascinated millions of television viewers around the world. And that is just what MSC is—control center. No launch vehicles ever lift off here; that is the job of Cape Kennedy. But once a particular manned launch vehicle has cleared the pad, the Cape's job is done, so to speak. From there until splashdown, the astronuats' primary communications link with earth is MSC. Houston Control is rapidly extending its boundaries with the advancement of Project Apollo. Just outside the eastern perime-

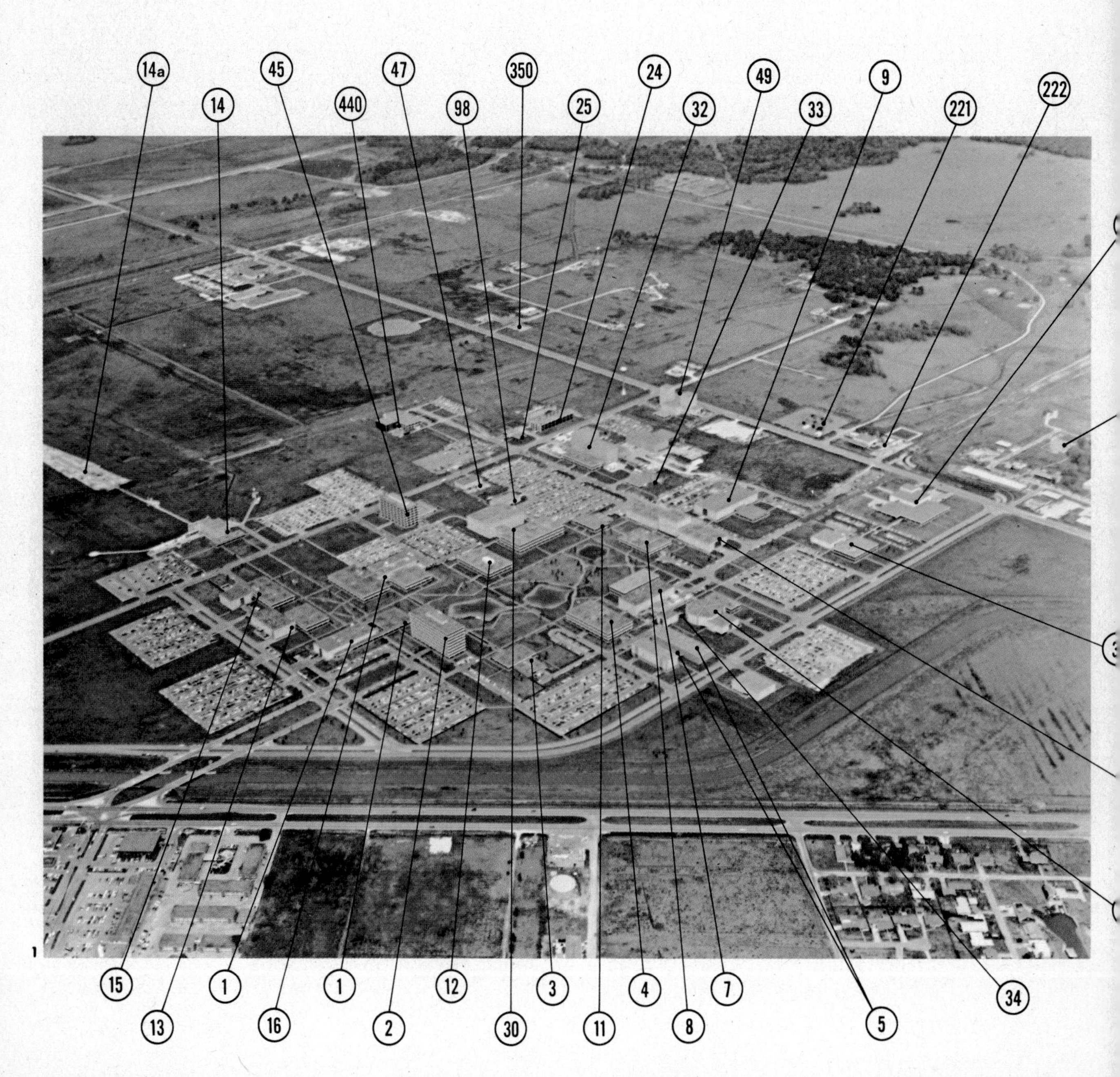

ter is the Lunar Science Institute (**2**), formerly the mansion home of the late rancher James "Diamond Jim" West. Its sumptuous interior (**3**) still reflects gracious living of an earlier era. Of special interest to visitors and tourists is Building 1 which houses an 800-seat auditorium, in which NASA films are shown, and the museum (**4**) containing a superb collection of exhibits and artifacts from America's space exploits. Exhibits are constantly updated to keep pace with the latest missions. The rear of Building 1 is a separate facility for members of the press. Representatives of the media gather at preappointed times (**5**) to get the latest information, as in this briefing by Flight Director Glynn S. Lunney. MSC is the astronauts' professional home and training ground. Here they not only undergo their "boot camp," but also tackle various training and testing programs to qualify them for specific mission operations. To this end, MSC contains extensive facilities for the training of the men and the development and proving of equipment. The astronauts are by no means just "along for the ride" during a mission. They form a strong connecting link between research and development and operations. Manned Spacecraft Center's construction was begun late in 1961, under the management of the Army Corps of Engineers. Prior to this time, NASA's Space Task Group had been based at Langley Research Center in Virginia. For a time, Ames Research Center in California was the prospective site, but was shifted at President Kennedy's directive.

3

MANNED SPACEFLIGHT CENTER

1. Space museum & auditorium
2. Project management building
3. Central cafeteria & souvenir shop
4. Flight crew support building
5. Mission simulation & training facility
7. Life systems laboratory
8. Technical services office
9. Technical services facility
10. Technical services shop
11. Branch cafeteria
12. Central data office
13. Systems evaluation laboratory
14. Anechoic chamber test facility
4A. Antenna test range
15. Instrumentation & electronic systems lab
24. Central heating & cooling plant
25. Fire station
29. Flight acceleration facility
30. Mission control center
31. Lunar mission & space exploration facility
32. Space environment simulation lab
33. Ultra-high-vacuum space chamber facility
34. Flight acceleration motor generator building
37. Lunar receiving laboratory
45. Project engineering facility
47. Southwestern bell telephone building
48. Emergency power building
49. Vibration & acoustic test facility
21. Electrical substation
22. Atmospheric reentry material & structures evaluation facility
70. Solar telescope facility
50. Thermochemical test facility
40. Electric systems compatability facility

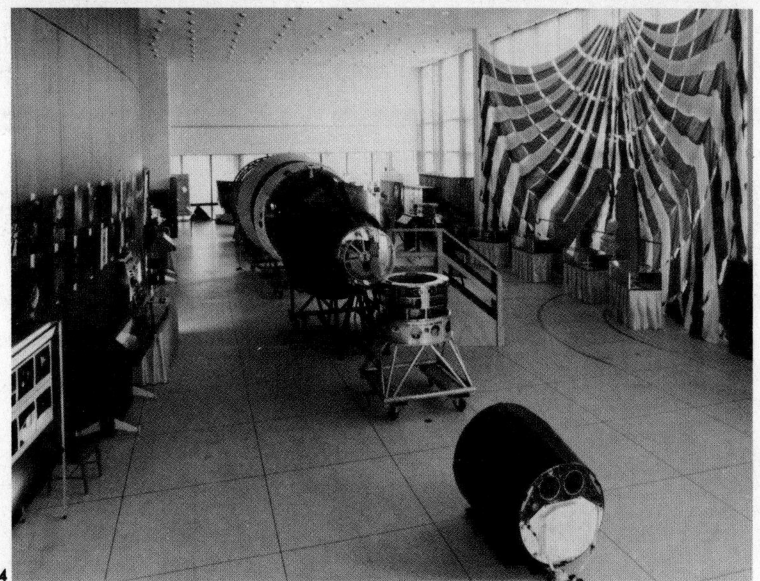

4

5

SPACECRAFT CENTER

One of the most important facets of the astronauts' training program is physical conditioning. At MSC, a complete gym has been provided. Here, Joe S. Garino, Physical Training Specialist, coaches astronaut Joe Kerwin (**1**). At left is astronaut M. Scott Carpenter. Garino designs an individual workout program for each of MSC's space flight and aircraft crews, depending on each man's needs. Training for Apollo, astronauts Stuart Roosa and John Swigert are inside the Dynamic Crew Simulator in Building 5 (**2**). Projectors cast life-like scenes of the approach and descent to the lunar surface through windows of the "module." The complete Apollo Mission Simulator (**3**) is seen from the operator's station. The entire set-up weighs nearly 40 tons and stands some 30 feet in height. Future Apollo crews sit within the multifaceted cabin in the background; technicians at the consoles work with the astronauts simulating mission procedures, or against them creating unexpected "emergencies" that could occur in reality. The

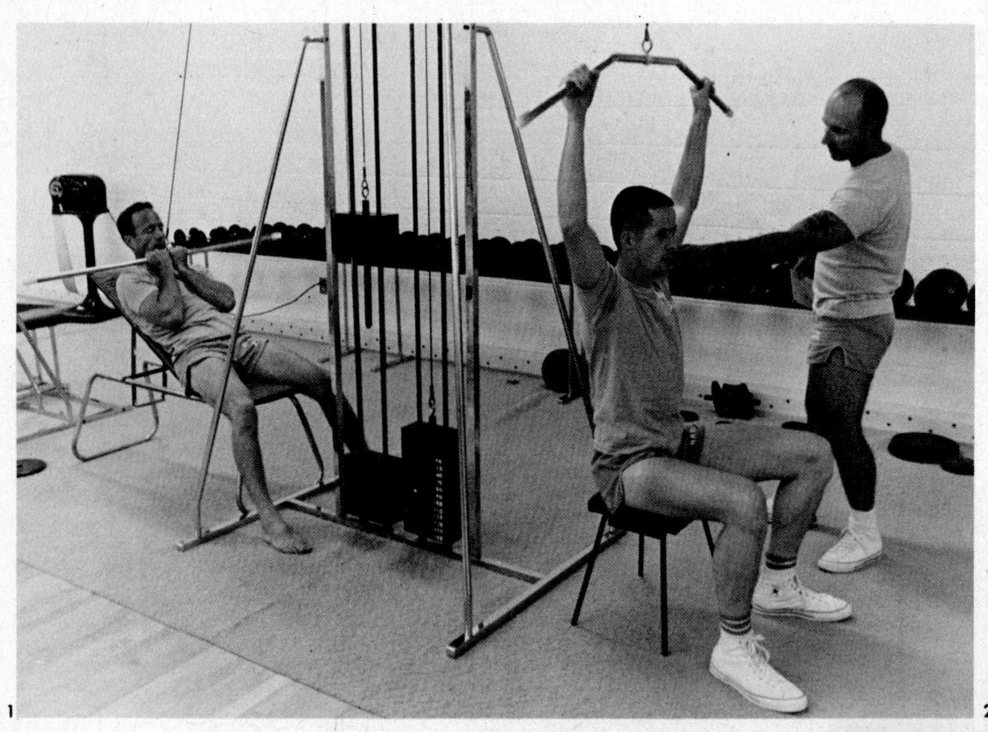

1

2

5

6

Univac 1108II (**4**) is a third generation system, and is one of four such computers installed in Building 12, the Computation and Analysis Division Facility. The computers are used to support both the scientific and management environments of MSC. Building 16 houses the Electronic Scene Generator (**5**). Technicians monitor this vital system which is essentially a segment of the sophisticated training system at Manned Spacecraft Center. The scene generator is being programmed for the future Space Shuttle operations, which will include recoverable spaceships that will land on runways in the same manner as an aircraft. A view of a landing strip scene (**6**) generated by the simulator will give future astronauts life-like approach and touchdown scenes while training for the Space Shuttle. The Shuttle, when realized, will permit untrained or semi-trained scientific and other research personnel and their cargo to be taken into space by a crew of astronauts to a permanent space station, and subsequently returned to earth. MSC works in the future as well as meeting the research and operations necessary for current programs. Building 25 is MSC's fire station, (**7**) fully manned and equipped facility which stands ready 24 hours in case of fire or other disaster. The fire station serves to illustrate the fact that MSC, quite distant from both Houston and Galveston, has become a small, self-supporting city in itself with all facilities normally considered municipal functions. Its "police" are its security guards, and its "mayor" is the Director.

3

4

SPACECRAFT CENTER

There are many strange scenes awaiting the uninitiated person at Manned Spacecraft Center, but none could be more commanding than the anechoic chamber (**1**). Located in the Instrumentation and Electronic Systems Division's Building 14, this huge room, or cell, is completely lined with foam pyramids that absorb stray radiation during spacecraft antenna pattern testing. The word "anechoic" means literally "against echo," and it's a very quiet place. The maneuverable tower is part of a test rig for lunar surface antenna studies. Here is a familiar looking object to the person watching Project Apollo (**2**). This is a full scale mock-up of the Lunar Module's ascent stage. It is being mounted atop the Instrumentation and Electronic Systems Division's Boresight Range Control Building for an extremely accurate alignment of the LM's rendezvous radar system. The pedestal is a finely adjustable three-axis positioner. Located in the Flight Acceleration Facility is this large centrifuge (**3**). Astronauts spinning in the globu-

1

2

5

lar cell at the end of the 50-foot arm can become familiarized with the G-forces encountered during a launch or re-entry. The centrifuge was built for Apollo training purposes. The interior of Building 32 (**4**) appears to be filled with machinery and piping for a refinery, but actually it is the Space Environment Simulation Laboratory. Large chambers are used into which spacecraft can be placed and subjected to the various conditions they will encounter when outside the earth's atmosphere. Simultaneously, engineers can observe and check out the craft's systems under deep space conditions. One of the most important buildings at MSC is the Lunar Receiving Laboratory (**5**). The LRL is a complete self-contained analytical laboratory where equipment and specimens brought back from the moon will receive their initial scientific examination. Here, also, the crew returning from their lunar voyage will live under quarantine conditions for a short period until doctors are satisfied that they have not brought back any alien bacteria. The Lunar Receiving Lab has been built specifically for Project Apollo. Three astronauts participate in Apollo water egress training in Building 260 (**6**). Climbing out of the Apollo Command Module Trainer is Tom Stafford, while already in the life raft are Eugene Cernan (back to camera) and John Young. They are the prime crew for a future Apollo mission. Every phase of Apollo is rehearsed at MSC. Through simulators, astronauts will have been to the moon and back many times before the actual landing.

4

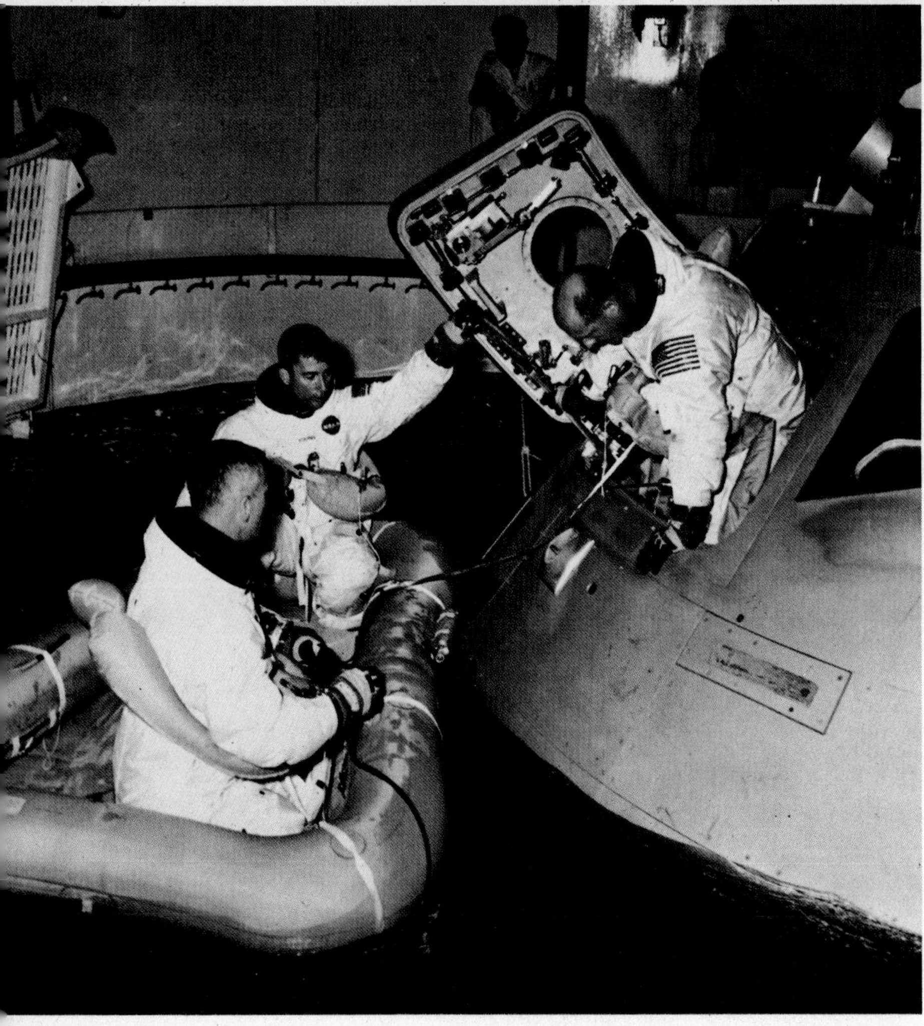

SPACECRAFT CENTER

Nestled among a grove of trees in the south corner of MSC is Building 270, the Solar Telescope Facility (**1**). Within the small domed building is a special telescope (**2**) that is part of a group of such instruments. Together, they form the Solar Particle Alert Network (SPAN). During future deep space missions, astronauts will be outside the earth's protective magnetosphere, and subject to solar radiation of varying intensities. The solar observatories will permit a 24-hour watch for intense solar flaring or "storms" which would expose the crew to stronger radiation. Mission Control will be able to warn the astronauts immediately of solar flare-ups, so that they can take protective countermeasures. During a mission, the Operations Control Room in Building 30 (**3**) is the activity center. The astronauts are never really alone in space; all systems aboard their spacecraft, including their own biological functions, are monitored from this room continuously. A worldwide tracking and communications network focalizes all its ca-

1

4

pabilities on Mission Control, keeping them informed of the mission's status. This is very critical, because otherwise there would be blackout periods while the earth rotated. Controllers work in shifts, each man having a particular area to monitor. He is in constant communication with his fellow controllers. Should a crew become incapacitated in space, or a control failure occur, Mission Control can assume functional operation of the spacecraft and return it to earth. A man with an important job is Dr. Charles Berry (**4**). As Chief, Center Medical Programs, he is the "country doctor" for the astronauts. Fully conversant with their biophysical condition on earth, he constantly monitors their bodily functions at his console while they are in space. A strip chart at his station constantly prints out the crew's heartbeat, respiration rate and other functions. Paul Haney (**5**) is the MSC Public Affairs Officer. At his console, he is constantly monitoring the progress of a mission, and it is his responsibility to keep the media and the public informed of the mission's progress. He is among the small group of controllers who are allowed to communicate directly with the astronauts while on a mission; normally this is the function of the Capsule Communicator (Capcom). Each controller has a mission evolution or flight plan book before him, allowing him to know instantly at which point a particular event is to be initiated. Christopher C. Kraft, Jr. (**6**) is the Assistant Director for Flight Operations. The boss in Building 30, he is the senior mission controller.

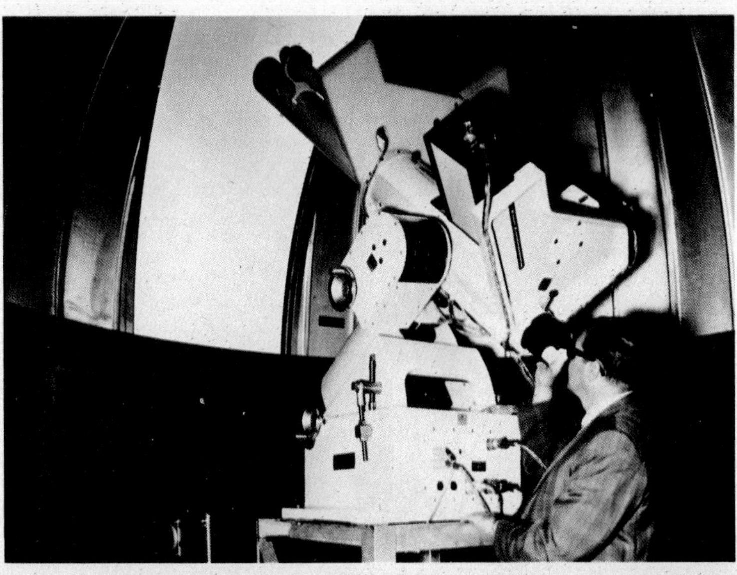

2

3

6

There is extraordinary concern at Star Town for the mission of Voskhod 2. A man is going to step from his spacecraft into the nothing beyond earth's atmosphere. How will he react, strung on a slim tether to the capsule, whirling through the void a hundred miles from home? There is absolutely no frame of reference for him to engage: no pressure against his body, no sound, no visual enclosure save the disarray of stars. It is known the solar glare could blind and cripple him in seconds, but this alien world must be confronted; men must train to become walkers in space, not just pilots. Whether repairing a damaged craft or building space stations, extra-vehicular activity is inseparable from man's push beyond earth. Chosen for this daring exploit is Aleksei Leonov, a 30-year-old cosmonaut from Siberia who is fond of painting. He and his companion for the flight, Pavel Belyayev, undergo training in the thermal decompression chamber and the isolation room. Psychologists pore over the sketches Leonov draws, and note he

LEONOV'S WALK IN SPACE

Another space first for the Russians

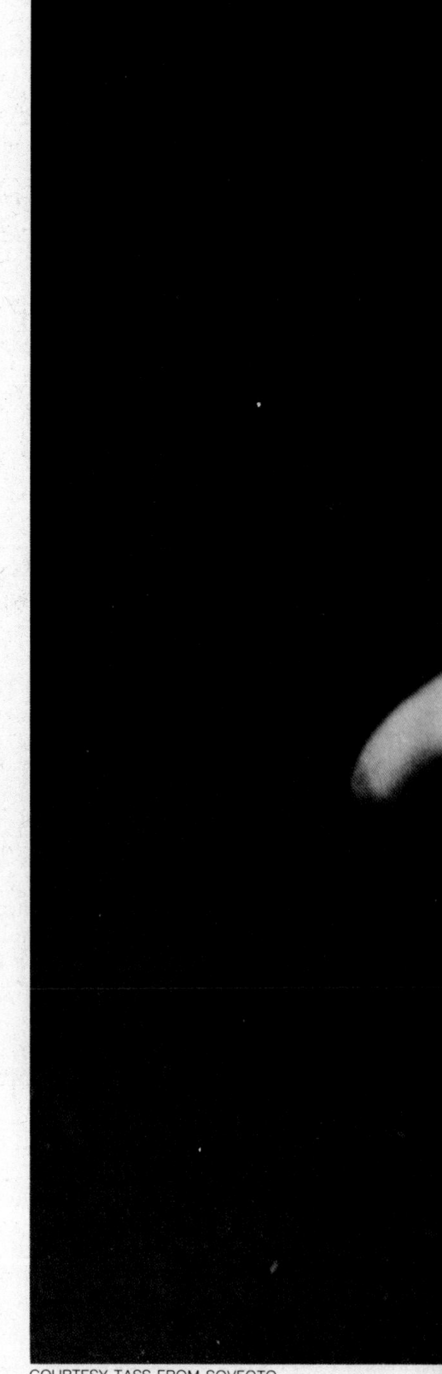

COURTESY TASS FROM SOVFOTO

seems to hold up well. After weeks in isolation, they immediately send him up in a plane, from which he is ejected; the doctors observe that he also handles this abrupt transition in spatial orientations satisfactorily. On March 18, 1965, at 10:00 AM, Voskhod 2 is launched into an orbit roughly 108.4 miles at perigee and 311 miles at apogee, making a revolution every 90 minutes. In the second of these Leonov dons a backpack containing his EVA life support system, and with the help of his partner climbs into the lock chamber. And then the hatch is opened to brilliant light, the sunstorm of deep space. Leonov moves to the edge of the hatch, removes the cover from a movie camera attached to the capsule's skin; he lets go, comes back, pushes away again—this time to the length of his 17½-foot tether. For ten minutes he is free in space; no man out of all the billions has stood as he does. But then it is time to return, for the oxygen supply is limited. He brings the camera with him, but in the cramped space of the lock chamber hatch it wedges with him. Struggling to get in, he tires, the oxygen is being depleted; finally, though, he pushes the camera in and follows after it. He is safe with his history. After another orbit he and Belyayev take Voskhod 2 down manually, landing in a forest of the Ural mountains at 12:02 PM, March 19. Because of the manual re-entry they are far from their intended landing site. So these men who've given the world its first space walk build a campfire, pitch a tent, and wait for the world to hear.

The primary goal of our space program is to put a man on the moon by the end of the 1960's. Our first step towards this objective was Project Mercury, and from that series of missions we have learned a great deal. Of equal importance, we have also found out that there is a great amount of information and skills which NASA and the aerospace industry have yet to develop and integrate into space science and technology in order to achieve the moon landing. Project Gemini has been conceived as a stepping-stone between the essential information gained during Mercury and the ultimate goal, Project Apollo. A more suitable name for this interim project could not have been found. The Gemini spacecraft is primarily a highly uprated version of the Mercury capsule, designed to accomodate two astronauts; like the twin stars located in the constellation Gemini, Castor and Pollux, the two astronauts will be a brotherly team in space. And this is one of the two primary objectives of Project Gemini: put two men into a spacecraft, and as-

PROJECT GEMINI

An introduction to America's two-man spacecraft

1

2

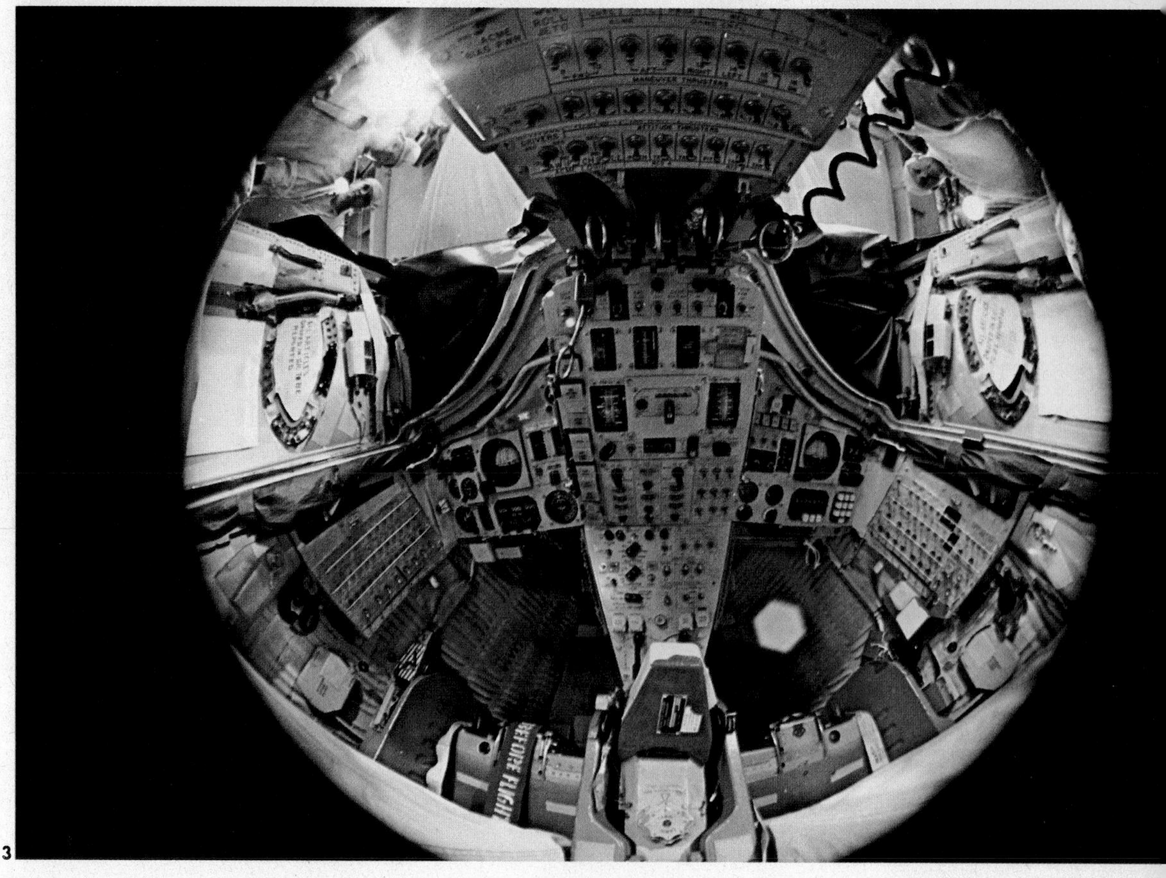

3

sign them active flight and experimental tasks on voyages which will extend far longer than any Mercury mission. In providing for two men and more sophisticated equipment, the Gemini capsule, while resembling the Mercury capsule, is much larger (**4**). The second objective of Gemini will be to study and perform in-space "docking" or linking up of two space vehicles (**1**). This is in anticipation of future Apollo missions in which two spacecraft will make the lunar voyage linked together. The workhorse booster of Project Gemini will be the Titan II, (**2**) based on General Dynamics' two-stage USAF ICBM. For Gemini, flight control and malfunction detection systems will be added to this rocket. In Gemini configuration, the launch vehicle stands 108 feet in height (**7**). Returning to the Gemini capsule (**6**), it can be seen that entry and exit of the astronauts is through a pair of hatches. These also serve as escape hatches; in case of a mission abort on the launch pad or during lift-off, the astronauts will be ejected out these openings in their couches in much the same manner as a jet fighter pilot. The Gemini cockpit (**3**) resembles that of a modern aircraft. McDonnell Aircraft, builder of the Mercury capsules, remains the prime contractor for the Gemini spacecraft. At the Cape (**5**) NASA has constructed two new launch pads for Gemini, numbers 14 and 14A. The first of these is for the Gemini-Titan rockets; 14A will be the launch point for the unmanned Atlas-Agena target vehicles, with which the Gemini spacecraft will rendezvous and dock.

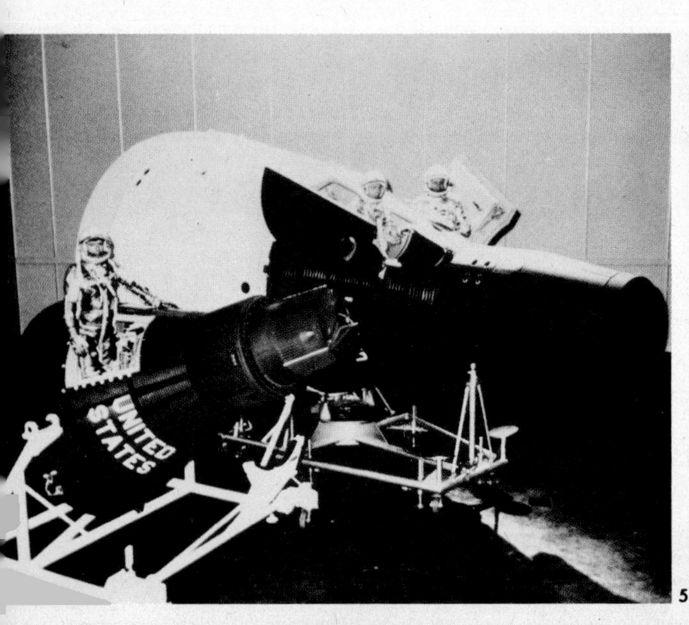

5

7

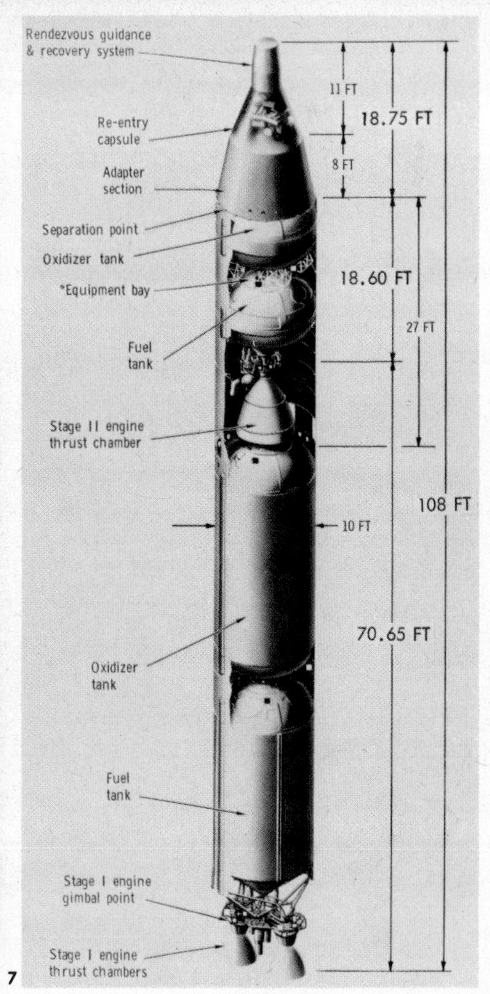

PROJECT GEMINI

One reason for the increased size of the Gemini capsule is its provisions for advanced maneuvering propulsion systems, necessary for any docking attempts. The large white section at the rear of the craft (**1**) houses the Orbital Attitude and Maneuvering System (OAMS); its terminology describes its function perfectly. Small reaction motors placed around the outer section will allow the astronauts to make very precise maneuvers in space. At the forward end of the capsule is a cylindrical section which houses the Re-Entry Control System (RCS). Small thrusters here ensure correct attitude control of the craft during re-entry, a phase that was always something of a "sweaty-palms-of-the-hands" phase of the Mercury flights. The OAMS is jettisonable, and will be discarded prior to re-entry. The two sections of the capsule are shown separated in this rendering (**2**), along with the location of the instrument equipment section. A third cutaway (**3**) shows that the two astronauts sit at a slight diverging angle

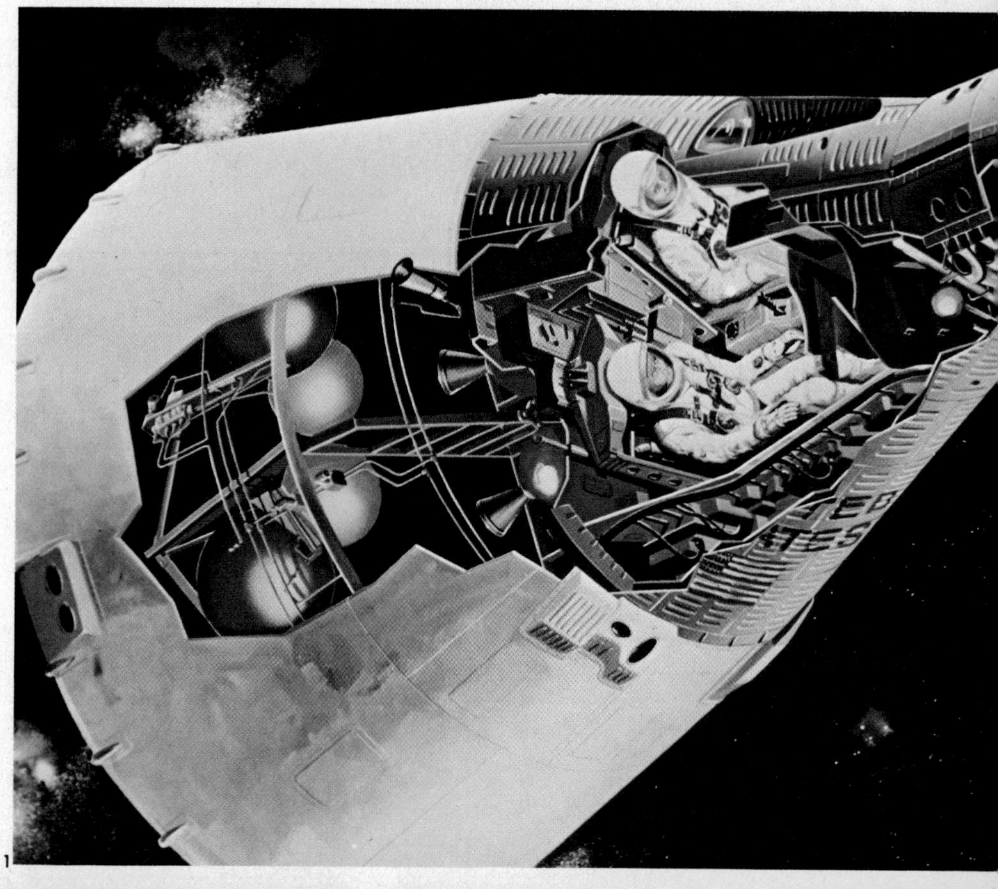

1

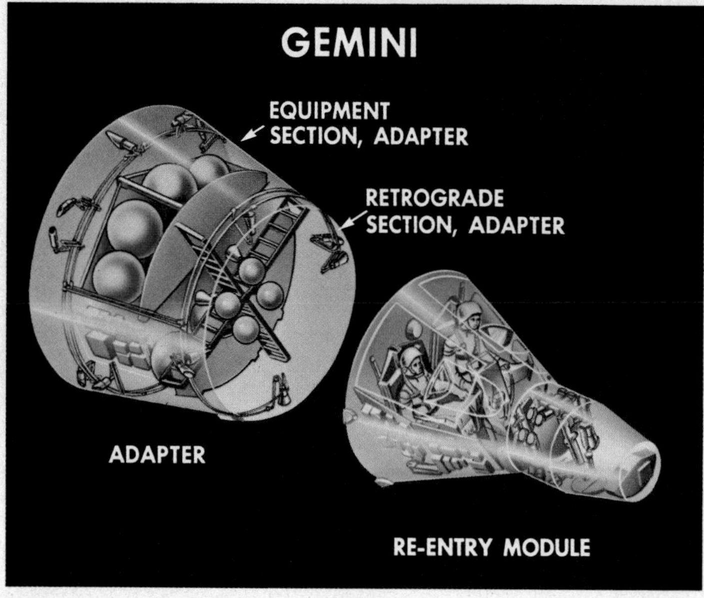

2

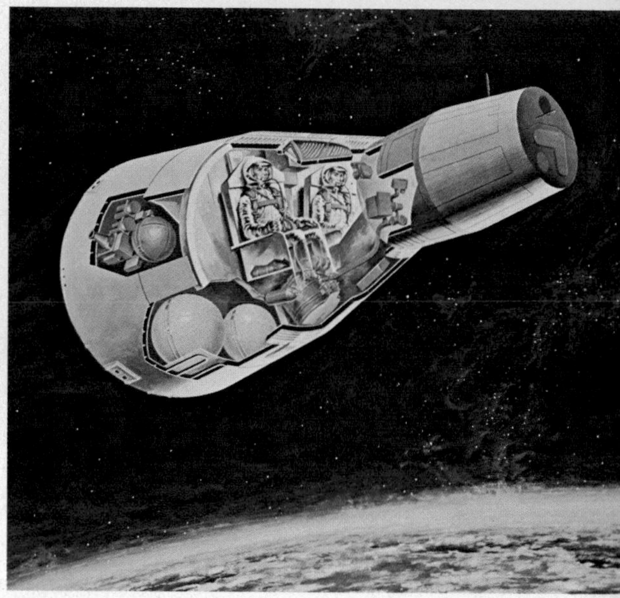

3

from each other. Also shown in position are the two helium tanks, used to pressurize the reaction engines' fuel tanks. Assembly of the Gemini capsules (**4**) is carried out at McDonnell's St. Louis plant. The contract with NASA calls for 12 orbitable spacecraft, along with a number of mockup "boiler-plate" capsules for various tests. Positioning of vital equipment has been made with an eye towards easy checkout, integration and on-the-spot repair. This had been a big problem at times during Mercury; to replace one switch, entire components first had to be removed, calling for another complete checkout of each one when they were reinstalled. Astronaut Jim Lovell (**5**) is seen in the pilot's seat of a Gemini capsule during a checkout session. Here (**6**) we look into the cockpit of a capsule under construction. The heat shielding is similar to that on the Mercury capsules, consisting of an outer shell with ribs or "shingles", made from a heat-resistant nickel alloy. Red shrouds cover the two portholes on the hatch lids. Technicians (**7**) mill the Gemini capsule's main re-entry heat shield with a large diamond-coated mill. The slightly convex face is coated with an ablative material, which literally burns off at a controlled rate. This causes rapid dissipation of the searing heat generated during re-entry, as the ablative material vaporizes into the atmosphere and takes the heat with it. The Gemini spacecraft will weigh about 7000 pounds at launch, and stands 18 feet, five inches in height. The splashdown weight is about 4700 lbs after the OAMS jettison.

4

5

7

PROJECT GEMINI

For astronaut training and evaluation of the spacecraft's various systems, a functional mockup has been installed at the Cape (**1**). A similar device was utilized prior to and during the Mercury program, and proved its worth—astronauts and ground control personnel can "fly" a number of missions without ever leaving the room. At the Naval Weapons Testing Center in China Lake, California, technicians prepare a boilerplate Gemini capsule for a ride on a rocket sled (**2**). Its purpose will be to test the operation of the twin ejection seats under simulated high-velocity abort conditions. Instrumented dummies, not men, will ride out the run. One man is adjusting a 16mm movie camera to record the sequence. Meanwhile, utilizing another boilerplate capsule, NASA studies the splashdown situation. Certain designers feel that a large para-kite (**3**) will answer better than regular parachutes for the final ride back to earth. Dropping dummy capsules from cargo aircraft at high altitude allows a helicopter to follow the sequence. A

1

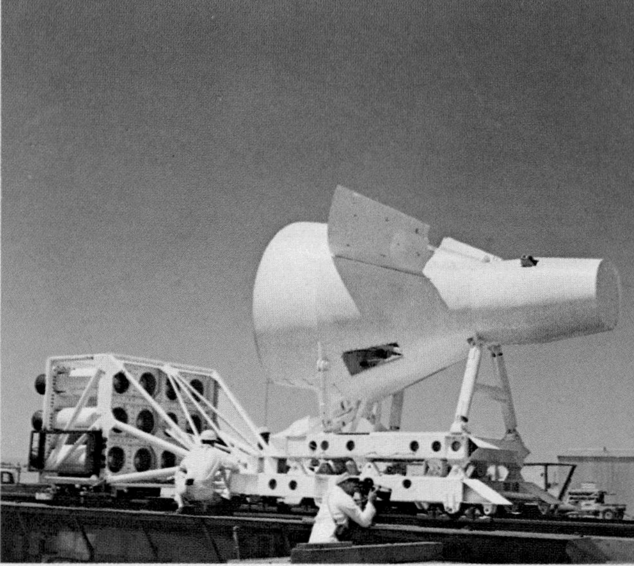
2

3

para-kite will allow the capsule to be maneuvered during descent, permitting better teamwork between the astronauts and the recovery forces. Further, it will reduce landing impact, and with a tricycle-type landing gear, may even make ground touchdowns feasible. Simultaneously, a modified Mercury parachute landing system is being developed for Gemini, also designed to lower impact at splashdown. Each Gemini capsule will carry a complete emergency survival kit (**4**), should the crew be separated from the recovery forces for an extended period after splashdown. McDonnell technicians (**5**) perform a propellant indicator calibration during a spacecraft systems test at the Cape. Modifying the Titan II missile for manned purposes has proved a challenge similar in some ways to the same problems that surrounded the Atlas during Project Mercury. Today, the Atlas has become the "workhorse", while the Titan has become the problem child. Resting prone on its erection tower (**6**), the Titan is an impressive piece of rocketry. Its two Aerojet General LR-87AJ7 engines produce a total of 430,000 pounds of thrust, while the second stage LR-91AJ7 engine produces 100,000 pounds thrust. These engines are of the liquid-propellant type. Astronaut Edwin Aldrin practices EVA procedures in a large tank of water (**7**). This condition approximates zero gravity in space. He is placing his feet in a special footplate located on the adaptor section of the Gemini spacecraft. Learning to work in space is an important phase of Gemini.

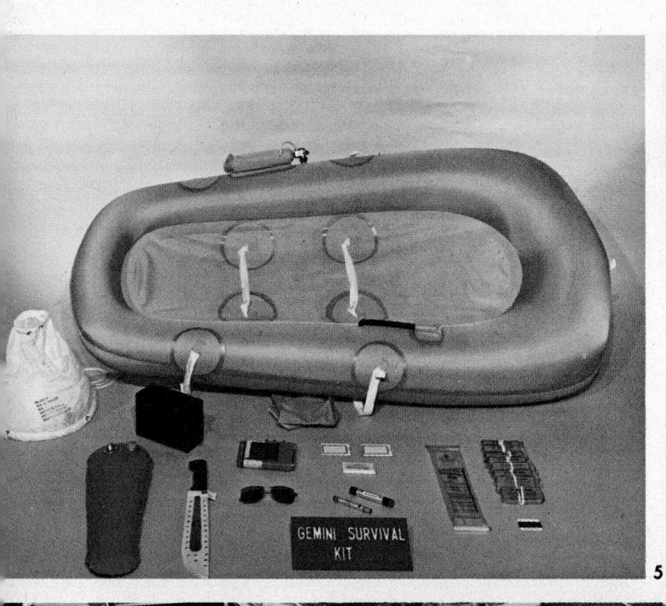

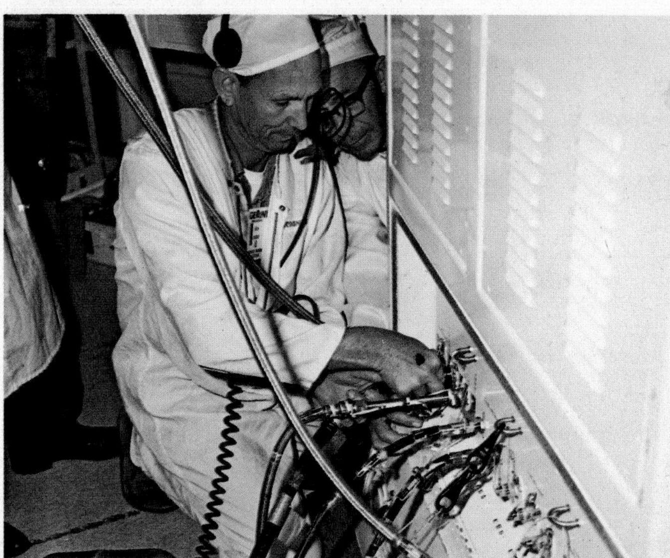

5

7

PROJECT GEMINI

In this simulated scene, a command pilot in a Gemini capsule (**1**) prepares to rendezvous with an unmanned Agena target vehicle. To the right corner of the cabin window is the cone-shaped docking drogue, which is designed to mate with the Gemini's forward section. The Agena is powered by a restartable Bell Aircraft Type 8096 engine of 16,000 pounds thrust. It is the second stage of the target rocket, boosted into orbit by the first-stage Atlas; the complete rocket is therefore called an Atlas-Agena. Advanced simulation procedures such as depicted here have placed our astronauts into a high state of readiness for actual flights. At the Cape (**2**), the launch vehicles are assembled at Launch Complex 19. Launch Complexes 14, 13 and 12 also will serve the Gemini program. Shrouded from weather, not for the sake of secrecy, the Titan II for the initial Gemini launch is assembled and checked out. The astronauts will stay on the ground for this flight, scheduled for April, 1964; it will serve only to evaluate the spacecraft's

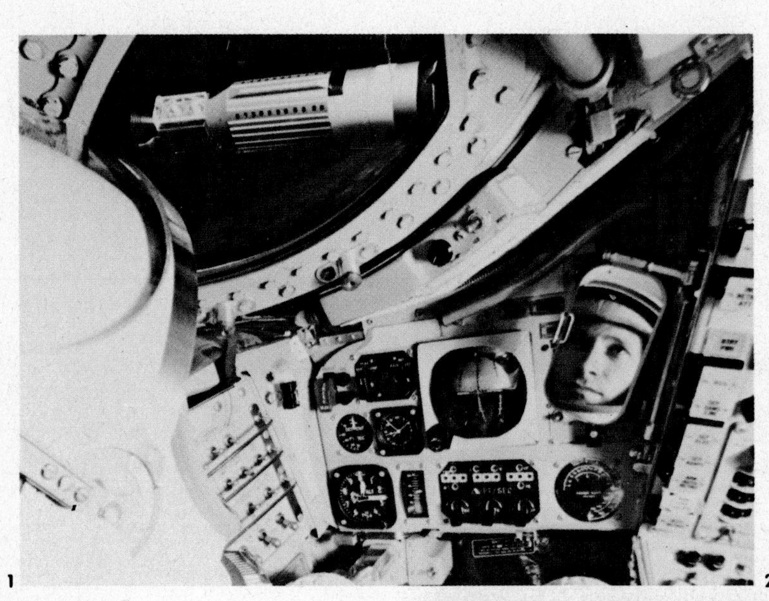

1

2

3

potential and the operations of the launch and tracking crews. A group of launch technicians (**3**) wear full protective suits during the handling and transfer of the Titan's fuel. Unsymmetrical dimethylhydrazine is as unstable as its chemical name is long! It has a tendency to explode when handled improperly. A beautiful sunrise (**4**) catches Launch Complex 19 swathed in its own artificial lighting during prelaunch checkout. At 11:00 AM EST, GT-1 lifts off from the Cape (**5**). The top of the Titan is an unmanned boilerplate capsule. The mission profile does not call for a separation of the dummy capsule from the second stage. All NASA wants to verify is the "compatibility" of all systems, and the ability of the Titan to put the spacecraft into orbit. Six minutes after launch, the second stage and capsule enter a successful orbital path, ranging in height between 100 and 205 nautical miles. In the launch control room (**6**) everything proceeds smoothly. Three orbits and four hours, 50 minutes later, the mission is called complete and a full success. However, NASA's Goddard Space Flight Center continues to track the craft until its orbit decays on April 12. The end comes for GT-1 over the southern Atlantic Ocean. Entering uncontrolled into the upper reaches of the atmosphere, it vaporizes in a fiery meteoric trail. The next launch for a Gemini vehicle will be suborbital—just a big ballistic arc downrange, this time to check the separation of the spacecraft capsule from the second-stage booster. Step by careful step, Gemini approaches readiness.

6

PROJECT GEMINI

The Titan II launch vehicle is stowed into a waiting Military Air Transport cargo plane for shipment to the Cape (**1**) in preparation for the launch of Gemini-Titan 2. NASA, in conjunction with the USAF, has now fired a total of 33 Titans from the Cape, thoroughly proving the missile's ability to function reliably. Packers (**2**) carefully fold and assemble the recovery parachute for the capsule. NASA has decided to go with the conventional parachute system of recovery, in place of the more exotic paraglider and parakite systems. By modifying the capsule so that it will splash down on its side, rather than its bottom as with Mercury capsules, the force of landing impact has been reduced to acceptable levels. Mission Control at Houston, Texas, is now working closely with the Cape in the launch program; once a rocket clears the pad, operation control of the mission will now switch to Houston. Gemini spacecraft #2 undergoes final checkouts. The capsule (**3**), fully instrumented, is hoisted to the top of the erection tower for

1

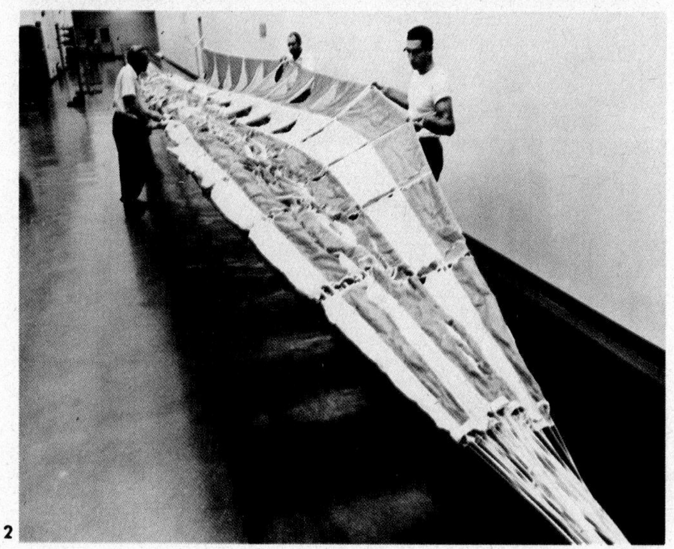

2

3

joining with the launch vehicle. In late November, the spacecraft completes what is termed a "Wet Mock Simulated Launch." This comprises a complete countdown, including fuel transfer. Crew suitup, ingress and egress procedures are tested during the countdown, using the prime crew for the upcoming manned Gemini-Titan 3 flight; they are astronauts Virgil Grissom and John Young. Launch countdown for GT-2 begins initially on December 9, 1964 and proceeds with fair regularity until one second after engine ignition, when a loss of hydraulic pressure in the launch vehicle causes automatic shutdown and temporary scrubbing of the launch. The hydraulic system is coupled to the Titan's guidance system, and governs gimballing action of the engine nozzles; without hydraulic pressure the Titan could not hold itself on course. On January 6, 1965, redesigned components for the hydraulic system are installed, and countdown is resumed. Launch of GT-2 (**4**) comes on January 19 at 9:04 AM EST, and the 18 minute, 16 second flight proves to be a success. Cameras inside the unmanned capsule record the flight (**5**) and the opening of the recovery parachute (**6**). Maximum altitude attained on this suborbital flight is 92.4 nautical miles. Six minutes and 54 seconds after launch, the retro-rockets fire. At 10:52 AM, the prime recovery ship, the carrier USS *Lake Champlain,* picks up the floating Gemini capsule. Curious sailors parade by the scorched but intact spacecraft on the flight deck (**7**). Manned Gemini is almost a reality.

4

5

6

7

GEMINI 3

And now, we are ready to begin again. Project Mercury proved that we could put a man into space and return him safely; Gemini will develop our abilities and further our potentials in space flight. It has been almost two years since NASA has put a man into space, but that time has been well spent in the design and development of new hardware for the tasks ahead. Sadly, that man who first called upon the country to devote its time, energies and money towards placing a man on the moon before 1970 will never see that momentous event. Just three short months after Gordon Cooper's 22-orbit flight in Mercury-Atlas 9, President Kennedy was felled by an assasin's bullet, prompting a wave of negativism and dissent that has shaken America in almost every quarter. But the echo of that call to peaceful strife still rings loud and clear to the members of NASA, now a powerful partner with our unmatched scientific and industrial potential. Gemini-Titan 3 will amount to a powerful restart of our thrust into space—manned space. The objectives of this mission are straightforward, like those of the early Mercury flights, but on surer footing. NASA will demonstrate manned orbital flight and the maneuvering capability of the .Gemini two-man spacecraft. Our uprated and expanded tracking network will receive an exercise in observation of the craft in orbit, as well as keeping the complex communications open at all times. Several secondary scientific experiments will be included, as is photography. The new space suit will receive its first in-space evaluation, as will all the human-oriented systems in the craft. Gemini 3 will place us on the stairway to the moon.

VIRGIL I. GRISSOM

COMMAND PILOT/ One of the original group of seven astronauts, Gus Grissom will be the first man to go into space for the second time. His first outing was in July, 1961 when he rode *Liberty Bell 7,* the last suborbital Mercury test, and almost drowned during the recovery. An Air Force Lt. Colonel, Grissom earned his wings in March, 1951. Subsequently, he has logged over 4,000 hours in the cockpits of jet aircraft. Born on April 3, 1926, in Mitchell, Indiana, he is a graduate of Purdue University and holds a BS degree in Mechanical Engineering. He is married to the former Betty Moore, from his hometown; they have two sons. Grissom is a member of the Society of Experimental Test Pilots.

JOHN W. YOUNG

PILOT/ A Navy pilot since 1953, John Young is among the second group of NASA astronauts selected in 1962. After flight training, he served in Fighter Squadron 103 at Jacksonville, Florida. Beyond that first tour of duty, he became a Naval test pilot, and has logged more than 5000 hours in military aircraft. Born in San Francisco on September 24, 1930, he earned a BS in Aeronautical Engineering with highest honors from Georgia Tech in 1952. Young likes to keep in shape and not only plays handball to do so, but also works out in the full pressure space suit to enhance his conditioning. He is a member of the American Institute of Aeronautics and Astronautics and the proud dad of two children.

NASA
GEMINI

GEMINI 3
Preparation

Pre-flight activities for Gemini-Titan 3 proceed at an expected pace—smoothly between the bumps. But the bumps are primarily "glitches" of an engineering nature that time will work out. Industry and NASA, relying on and utilising past experience, have learned to coordinate activities quite well. The U.S. Air Force, something of a silent partner in this seemingly civilian oriented atmosphere, has nonetheless made a tremendous contribution in aiding the man-rating and improvement of the Titan intercontinental missile for use as a launch vehicle. On January 4, 1965, Gemini spacecraft #3 is received at the Cape, and begins its series of pre-launch tests. NASA, specifically the Manned Spacecraft Center at the Cape, has decided to allow the Gemini manufacturer, McDonnell, to assume the greater burden of testing and checkout at its facilities in St. Louis, relieving NASA of certain post-production testing. This will result in flight-ready components being received at the Cape in a condition closer to launch status. The prime

1

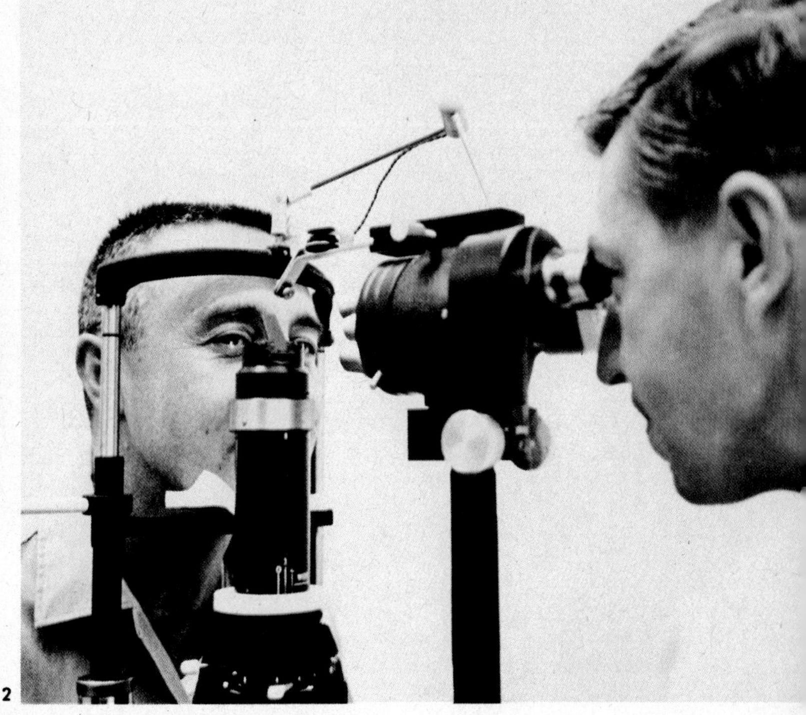
2

3

crew for Gemini 3 mission (**1**), Gus Grissom and John Young, discuss operations with MSC personnel. For Young, this will be his first flight into space, while Grissom has already spent 15 minutes, 37 seconds in suborbital trajectory in Mercury-Redstone 4 in 1961. It seems like a long time ago. A medical technician gives Grissom's eyes a close examination (**2**) during training. Young and Grissom are shown the intricacies of their spacecraft (**3**) in a "clean room" during its checkout. The technician is holding a fuel line to one of the capsule's maneuvering thrusters—a most critical piece of gear, because developing the ability to maneuver in space is one our main objectives in the Gemini program. Backup pilot Tom Stafford (**4**) at right, sits out a "duty watch" with a McDonnell engineer at the Cape during spacecraft qualifications. He is slated for a future Gemini mission, but staying on the ground does not leave any of the astronauts in a passive role, particularly the backup crew who train right along with the prime crew. Geronimo!! (**5**) John Young, practicing emergency escape procedures for an abort on the launch pad, makes an excited grimace for the camera. The cable on which he is sliding is connected to the launch tower, and in case of impending disaster at the site, the crew will be able to disembark from the capsule, attach themselves to a harness, and literally slide to safety on the ground. Complete with Grissom's and Young's pictures, the Titan launch vehicle (**6**) arrives at the Cape by air from Baltimore.

5

GEMINI 3
Pre-Launch

The crew shares an early morning breakfast with Deke Slayton (**1**, back to camera). Slayton, one of the first seven astronauts along with Gus Grissom, has since filled the role of Assistant Director for Flight Operations when it was discovered that he had a minor heart defect. John Young looks at the flight plan over Grissom's shoulder. All contractors have certified their products to be fully ready to assume the manned flight phase of Gemini. On January 15, the Titan's engines were fired on the ground during a final "end-to-end" check of the launch systems. All concerned are confident on this morning of March 23, 1965. Suited up (**2**), the crew moves to the van that will take them to Launch Complex 19. Into space with them will travel three scientific experiments. One will attempt to gain insight into the effects of zero gravity on living cells unconnected to a human organism. The second will measure effects of radiation and zero gravity on human white blood cells, while the third will eject water into the aura of ionized atmo-

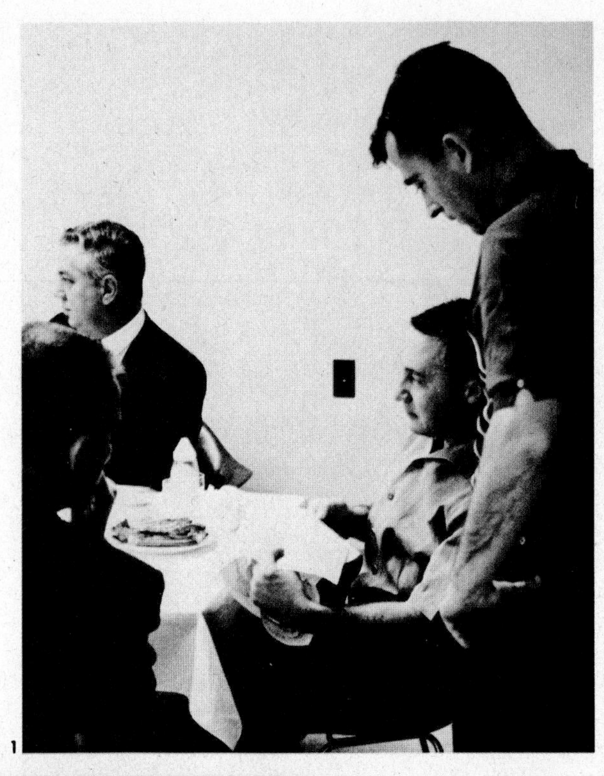

1

3

2

4

sphere that surrounds the capsule during re-entry. This experiment will attempt to offset the communications blackout which occurs during the superheated re-entry period of flight. As the countdown proceeds, the control room becomes alive with activity (**3**). An unperturbed launch controller in the foreground takes a moment to pull up his socks, however. In their suits (**4**), Grissom and Young look the part they are operating in. The shiny, silvery pressure suits are an improvement in design over the Mercury suit. With the help of launch personnel, they ease into their seats in the Gemini cockpit (**5**), Young on the left, Grissom on the right. A technician gives the right-hand hatch sealing gasket a final check. As the countdown proceeds, several minor technical holds are called, but none are of sufficient importance to delay or call off the launch. As the countdown proceeds to the launch, tension and attentiveness build throughout the Cape. Weather, always something which changes rapidly along the Florida coast, is on our side. Clear weather ensures the greatest safety level for the launch, and makes tracking and recovery much easier, especially if an early shutdown or abort of the mission is required. Among the expectant crowds watching Launch Complex 19 from a safe distance are the astronauts' wives (**6**). A little emotion beyond excitement on their part will be certainly justified! Suddenly, there is a blast of smoke from the launch pad, followed at a respectful interval by the ear-splitting roar of the Titan's engines.

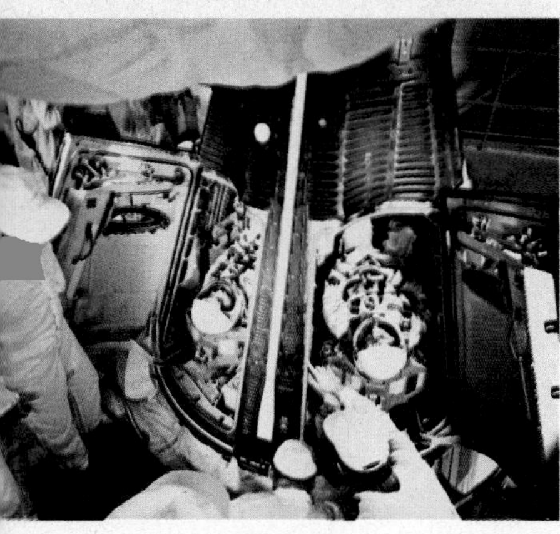

6

GEMINI 3
Launch/Mission

A resonant roar spreads across the flatlands of the Cape at 9:24 AM as the Titan II lifts off Pad 19 and into the sky (**1**). The second stage burns until its velocity has reached 17,500 miles per hour. Using the small thrusters, Grissom and Young boost themselves clear of the spent second stage. The orbit of Gemini 3 is almost circular; at its highest point it is 150 miles up, at its lowest point, 140 miles. Producing a variable thrust of between 25 and 100 pounds, the OAMS liquid-fuel maneuvering engines are one of the major advances of Gemini. They will allow the astronauts to make a calculated number of in-orbit changes in roll, pitch, yaw, velocity, and altitude; in short, to maneuver. An inflight set of photos (**2**) shows California's Imperial Valley and northern Mexico bathed in reddish sunlight. While entering the second orbit, the crew fires the nose thrusters. This reduces the spacecraft's speed by 66 feet per second, and a simultaneous attitude adjustment reduces their orbit to 100 by 107 miles. At two hours 20 minutes, another series of thruster maneuvers are performed, this time of a very precise nature. Velocity changes as small as two feet per second are dealt with, demonstrating the potential we have in the Gemini craft. The balance of the mission is consumed with more systems checkouts and the deployment of the three experiments. On this, the first Gemini flight, no attempts will be made towards any extra-vehicular activity. Tension builds again at the Cape (**3**) as the four-orbit mission nears its re-entry.

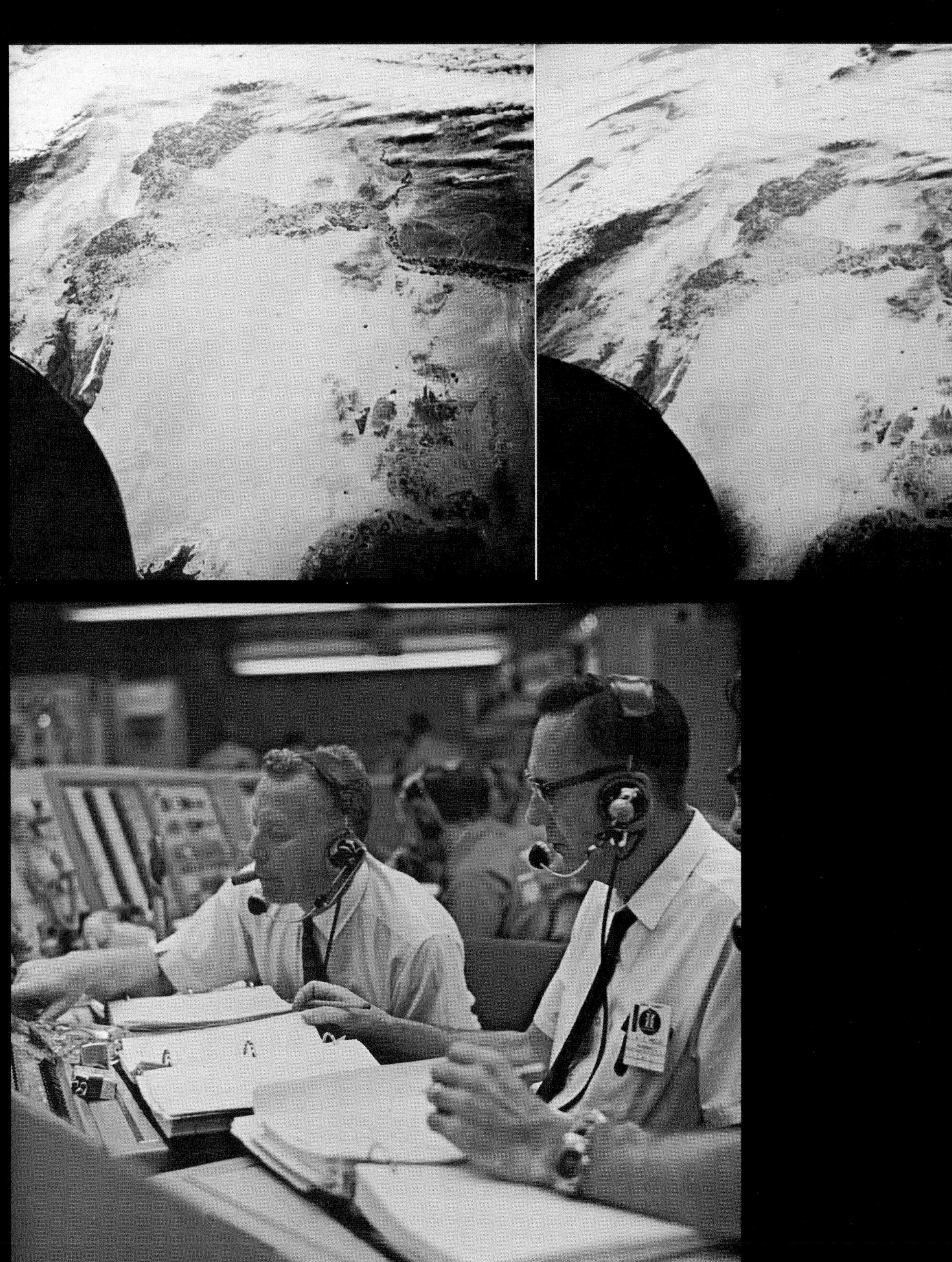

2

3

GEMINI 3
Splashdown

The four solid-propellant retro-rockets are ignited and burn themselves out, decaying the orbit of the capsule and causing it to fall towards the earth. Subsequently, they are jettisoned along with the equipment adaptor section. The OAMS thrusters are again utilised to properly align the spacecraft during re-entry. During the descent, the thrusters in the capsule's nose section will be used to help orient the capsule's descent, and bring it down close to the recovery fleet. Despite the added maneuverability of the Gemini capsule, splashdown occurs some 60 nautical miles away from the USS *Intrepid,* an irritating disappointment after an otherwise uneventful re-entry. After the four hour, 23 minute flight, two Sikorskys from the *Intrepid* hover over the floating capsule (**1**); helicopter No. 57 is the same aircraft that recovered Scott Carpenter after his three-orbit Mercury flight of May 24, 1962. After landing at 2:16 PM EST, Young and Grissom are to be hauled aboard the chopper and flown to the *Intrepid.* Frogmen (**2**) ensure the temporary

1

2

3

seaworthiness of Gemini 3. Unlike the Mercury capsule which floated on an inflatable bag surrounding its heat shield, the Gemini rides the waves on its side. The ablative heat shield appears very mottled after doing its work during re-entry. Their job over, the astronauts are hoisted aboard the helicopter shortly after 3:00 PM (**3**). Awaiting them are physicians who will give them thorough post-flight physicals, with an emphasis on sounding out any possible maladies resulting from weightlessness or radiation. Observations taken from the Mercury astronauts have reduced anxieties considerably in this area, however. The *Intrepid* moves in now, and at 5:03 PM, the sopping wet capsule is hoisted aboard and mounted on a stand on the flight deck (**4**). It too will receive numerous post-flight "physicals" from McDonnell "doctors". The design has proved its spaceworthiness beyond doubt, but there will always remain room for improving the breed, especially now with the post-flight recommendations of two seasoned men who have flown it some 81,000 miles. And for Gus Grissom and John Young awaits a proud nation (**5**), which despite ever-present skepticism from some quarters, is growing more aware of the fact that the challenge put forth by the late President Kennedy is being met. The successful flight of Gemini 3 has checked out the man-rating of the craft. From here NASA will advance towards more complex missions in the way to the moon. Next up is a spacewalk, and after that will come the critical rendezvous and docking.

4

5

GEMINI 4

The spectacular flight of Gemini 4 will herald an era of optimism and confidence in the American space effort. Jarred by the 10-minute space walk of Soviet cosmonaut Leonov, NASA springs into action. The flight of Gemini 3 successfully demonstrates that America has developed a maneuverable space vehicle, and the launch of Gemini 4 is moved up to more than a month ahead of its previous schedule. The main objectives are to demonstrate the performance of spacecraft systems in long-duration flight, and to evaluate the effects on man of lengthy exposure to space. But the secondary objectives are more spectacular than the primary ones. In addition to stationkeeping and rendezvous maneuvers, America will demonstrate its own extra-vehicular activity (EVA). This flight was not intended to have an EVA, but on May 27, 1965, nine days before the launch, it is announced that one of the astronauts, Ed White, will leave the orbiting Gemini to perform a space walk. This feat will be made possible by a new space suit, made especially for the hazards of moving about in space unprotected by the metal walls of a spacecraft. Fabricated by the David Clark company of Worcester, Massachusetts, it contains extra fabric layers to stop high-speed space particles called micrometeroids. White will not be able to go just anywhere he pleases, as he will remain attached to the Gemini spacecraft and its vital life-support functions by a 25-foot "umbilical cord", gold-coated to reflect solar heat away from its cool oxygen. The historic space walk is scheduled to last for twelve minutes. Jim McDivitt, without this suit, will remain inside the Gemini.

JAMES A. McDIVITT

COMMAND PILOT/ Born in Chicago on June 10, 1929, Jim McDivitt enlisted in the Air Force twenty-one years later, upon the completion of his junior college studies. As a jet pilot he flew 145 combat missions during the Korean conflict, winning three Distinguished Flying Crosses and five Air Medals. Intrigued by the concept of manned space flight, he returned to college and graduated from the University of Michigan with a B.S. degree in Aeronautical Engineering in 1959—first in a class of 607. He attended the Air Force Experimental Test Pilot School and became a test pilot at Edwards AFB, California, logging of 3000 flying hours. NASA picked him in September 1962. He has four young children.

EDWARD H. WHITE II

PILOT/ Born in San Antonio, Texas on November 14, 1930, Ed White took his first airplane ride at age 10, and his future was set then. An accomplished athlete and excellent student, he received an appointment to the U.S. Military Academy at West Point and graduated in 1952 with a B.S. in Military Science. He qualified as a jet pilot and served 3½ years in Germany. To fulfill his desire to become involved in the space program, he returned to college and received a Master of Science degree in Aeronautical Engineering from the University of Michigan in 1959. Another alumnus of the Air Force Test Pilot School and Edwards AFB, he became an astronaut at the same time as McDivitt.

NASA
GEMINI

GEMINI 4
Pre-launch / Launch

McDivitt and White (**1**) spend days in Gemini mission simulator #2 utilizing a planetarium to study orbits, constellations, and other stellar phenomena. Gemini 4, mounted high atop the Titan booster, also undergoes last minute preparations (**2**). On May 27, 1965 it is announced that an EVA will be attempted. Carefully, Ed White and a suit technician make a final check of the thirty-four pound, twenty-two layered spacesuit (**3**). The EVA suit is adapted from the normal flight suit designed for the Gemini program, but it weighs some 10 pounds more since it contains an additional thick layer of felt to absorb micrometeoroids. White has rehearsed the art of space floating in an altitude chamber for more than 60 hours and during the EVA he will be attached to the spacecraft and its life-support systems by a one-inch-diameter "umbilical cord". This is 25 feet long, weighs 9 pounds and is wrapped with a gold-coated plastic tape. The astronauts head for their space craft (**4**), and at 10:16 AM on June 3, Gemini 4 blasts off (**5**).

1

2

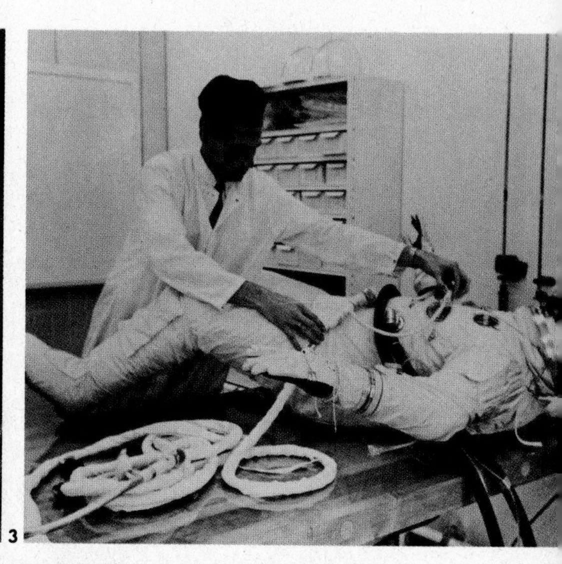
3

4

UNITED STATES

GEMINI 4 Mission

Man in space—135 miles up! Both astronauts are in their pressurized space suits and the cabin has been depressurized. The hatch is open and Ed White drifts away from Gemini 4 to start the longest space walk to date (**1**). The spacecraft is passing just north of Hawaii as he propels himself away from the hatch with the Hand-Held Maneuvering Unit (nicknamed the "ZOT" gun). This 7-pound unit (which also mounts a camera) squirts twin jets of oxygen when triggered, thus enabling the astronaut to propel himself around in space like a small rocket. This is the first unit of its type and the best maneuvering device to date, but it does not hold enough gas to be considered truly successful; however, it allows White to become the first self-propelled space man. Although White's oxygen, heat, suit pressure, and communications are controlled from the spacecraft by the umbilical cord (which also contains a tether lifeline), an emergency chest pack allows him to adjust his suit pressure and provides a backup oxygen

1

2

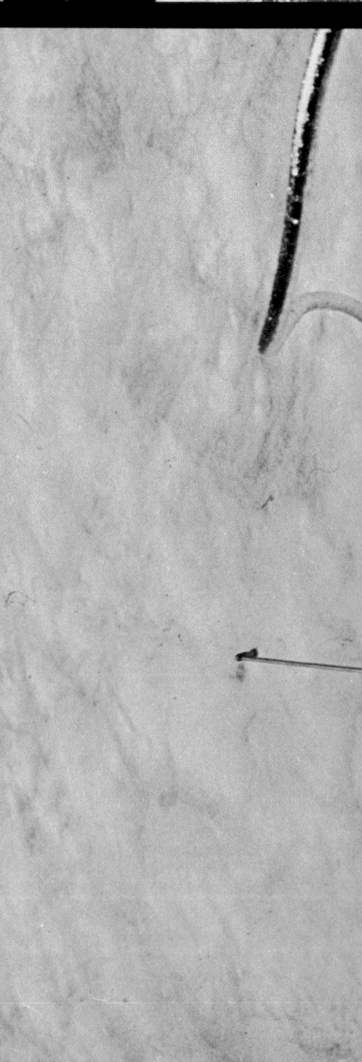
3

supply should there be some malfunction of the main source. Using the ZOT gun, White executes a slow roll (**2**). He spoke of the gun later, ". . . you can actually get from point A to B." Floating nonchalantly on his back high above a heavy cloud cover (**3**), Ed reports that he feels no disorientation or discomfort (contrary reports were given by Leonov). Coming in for a landing! (**4**) Astronaut White lands on the outside of the Gemini 4 capsule. He maneuvers himself all around the spacecraft and actually walks on top of it. Working his way forward, he has the bright idea of clinging to the Gemini's nose while McDivitt maneuvers the vehicle by firing the little thrust rockets. "But," he remarks, "I took one look at the stub antenna which was our radio connection back to earth and felt this wasn't any place to play around." The ZOT gun exhausts its oxygen supply quickly (**5**) and White maneuvers himself by the use of the tether alone, but finds it rather difficult to move in the direction he desires unless the umbilical is extended perpendicular to the vehicle's surface. When extended at an angle it causes him to roll when he pushes away from the spacecraft. By pulling on the umbilical he can bring himself back, and by exerting a firmer pull he can plant his feet on the vehicle's surface and walk around until the angle of the tether anchorage changes. On one of his transits he crosses McDivitt's window (**6**). "Hey," McDivitt radios, "you smeared up my windshield, you dirty dog. You see how it's all smeared up there?"

5

6

"Yep," White acknowledges as he tumbles slowly away (**1**). The olive-green tether stowage bag dangles in space (**2**) as Ed drifts out as far as the umbilical will allow, about 25 feet. "Hey," he calls, "I'm looking right down and it looks like we're coming up on the coast of California (**3**). I'm going in a slow rotation to the right. There is absolutely no disorientation associated with this thing. There's no difficulty recontacting the spacecraft. I'm very thankful in having the experience to be first . . ." Even if he experienced no disorientation, he did experience some euphoria and his pulse rate increased to 178. "I . . . felt red, white, and blue all over," he related later when being interviewed by *Life* magazine reporters. "Ed," McDivitt radios, "I don't even know exactly where we are but it looks like we're about over Texas. Hey, Gus," he calls to Virgil Grissom in Houston, "we're right over Houston (**4**)." "We're looking right down on Houston," says White, who was born in San Antonio. "Yeah," McDivitt agrees, "that's Galveston

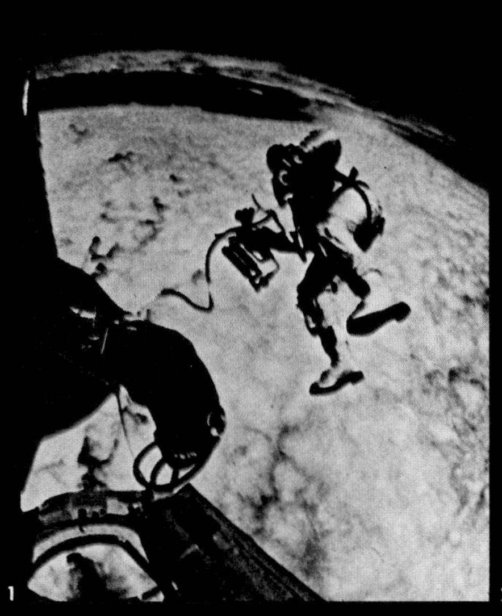
1

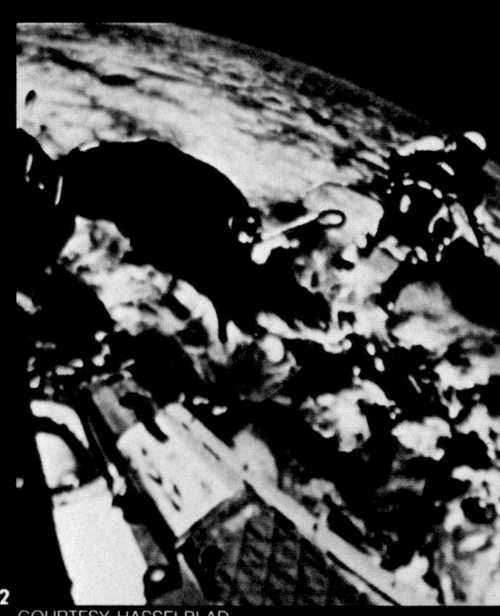
2
COURTESY HASSELBLAD

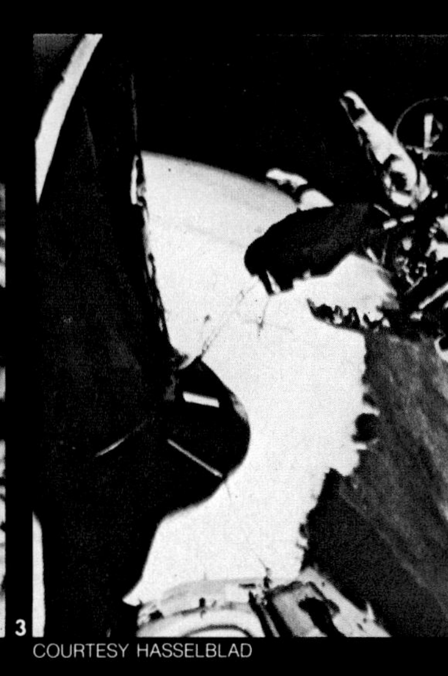
3
COURTESY HASSELBLAD

4
COURTESY HASSELBLAD

Bay down there." "I was taking some big steps," White says later in an interview for *National Geographic*, "the first on Hawaii, then California, Texas, Florida, and the last on the Bahamas and Bermuda (**5, 6**). The gun worked superbly; I just wish I had had more oxygen. Changing my position by pulling on the tether was easy, like pulling a trout, say a two or three-pounder, out of a stream on a light line." Minutes slip by quickly as the second man in space continues to cavort happily in slow motion—the proposed 12 minute sojourn stretches to 20 minutes. He radios to McDivitt, "The sun in space is not blinding. It's quite nice." He and McDivitt confer for a few minutes. During such conversations they cannot receive signals from Mission Control. McDivitt finally breaks off the conversation with White and calls Houston. They reply, "The flight director wants for White to get back in." The instruction is apparently not heard—Ed White has floated all the way across the United States. "Gemini 4, Gemini"—the voice belongs to Grissom—"get back in." "Okay," comes the response. Then White's voice is heard, "Hey, this is fun." "Well, come on back in," says McDivitt. "I don't want to come back to you," comes the response, "but I'm coming." Grissom radios, "You've got 4 minutes to Bermuda." "Come on, Ed. Let's get back in here before it gets dark," McDivitt's voice is urgent. Then White's voice is heard, "It's the saddest moment of my life." It takes both of them to close the hatch. Later, there's time for relaxation (**7**).

COURTESY HASSELBLAD

6

COURTESY HASSELBLAD

COURTESY HASSELBLAD

GEMINI 4
Splashdown

The remainder of the 4-day mission is relatively uneventful, and the balance of the flight is spent working on a series of medical and engineering experiments concerning the effect of weightlessness on the heart and the calcium content of the bones. Physical exercise with an elastic "bungee cord" is tried, and the amount and intensity of radiation encountered is measured and recorded. A new navigational aid is tested, and experiments are performed concerning the photographing of weather conditions and geological formations; over 200 photographs are taken. Food for the mission consists of freeze-dried, compressed, dehydrated meals which reportedly cost $255 per meal, and which provide each astronaut with 2,500 calories per day. The computer, which McDivitt requires to make a landing with pinpoint accuracy, malfunctions during the 48th orbit, and the splashdown is turned over to the Houston Control Center for the first time in the history of the space effort. As Gemini 4 approaches the sixty-third orbit, McDivitt

1

2

4

fires the retro-rockets that bring the capsule out of orbit and into a splashdown trajectory. Gemini 4 comes down in the Atlantic at 1:13 PM EDT on June 7, 1965—forty-two miles from the target point and ninety-seven hours and fifty-six minutes from blast-off. Later, aboard the aircraft carrier USS *Wasp,* McDivitt and White watch and talk while their spacecraft is hoisted aboard (**1**). Gemini 4 is carefully eased to the deck by a team of trained technicians wearing heat and radiation protective gloves (**2**). The spacemen and their craft had been in the water for 38 minutes before they were picked up by a Navy helicopter and deposited aboard the *Wasp.* They are found to be in excellent health and display no signs of vertigo, nausea, or weakness when examined by the Navy doctor (**3**), although White admits to feeling mildly seasick while bobbing around in the capsule awaiting the helicopter pickup. Awards are plentiful for the returned heroes: a congratulatory telephone message from President Johnson and an invitation to visit his ranch; later, the president announces that he has nominated them for the rank of lieutenant colonel; they visit Chicago and receive honorary Doctor of Science degrees; they are awarded NASA Distinguished Service Medals and astronaut's wings by the President (**4**). To dignitaries in Rome (**5**), McDivitt comments, "I did not see God looking in my space cabin window . . . but I could recognize His work in the stars . . . if you can be with God on earth, you can be with God in space."

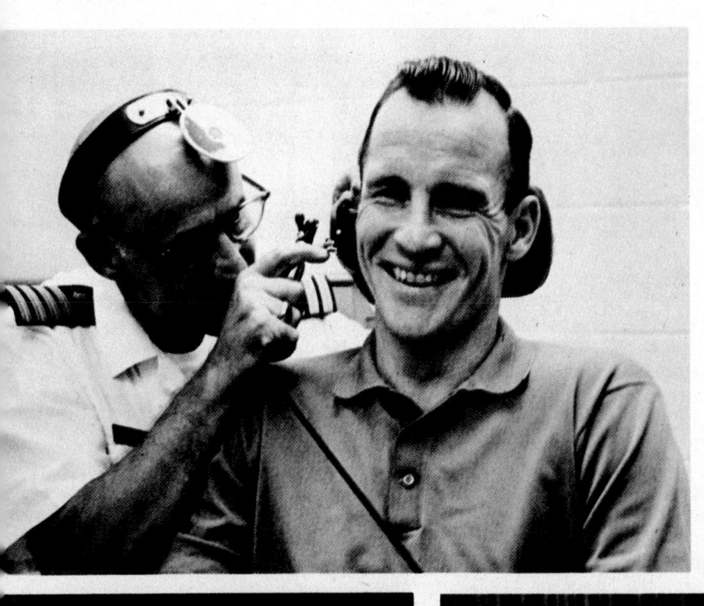

5

GEMINI 5

Continuing the general thrust of the Gemini series, with some important additions, the projected eight-day length of this mission (approximately the time required for a flight to the moon, surface exploration, and return to earth) will be the longest yet attempted by man. From August 21, 1965, through August 29, Gemini 5 will set a new record for man-hours in space by making 120 revolutions in 190 hours 55 minutes, and traveling 3,312,993 miles. Cooper will become the second man in the world to venture twice into space, in the process pushing the elapsed time of an individual in space to a record 225 hours 55 minutes. The back-up crew will be Neil A. Armstrong and Elliot M. See, Jr. Among the 17 experiments scheduled, the most important will be testing new equipment developed for rendevous and docking, the next major step toward the moon. Fuel cells capable of converting electrical energy from the reaction of oxygen and hydrogen, rather than traditional storage batteries, will be used for the first time. Mexico, East Africa, the Arabian Peninsula and Australia have been selected as priority objectives of terrain photography. It is expected that Gemini 5 will provide greater resolution than is obtainable from current TIROS weather satellites. Also scheduled is photography of the zodiacal light, a peculiar haze visible to astronauts just after sunset. The pioneering spirit of America's space exploration is symbolized by the covered wagon on the mission's insignia, the first one to be designed by the astronauts themselves; more than a century after the Conestoga wagons rolled west in search of a bright tomorrow, we still explore new worlds, seeking the same dream.

L. GORDON COOPER

COMMAND PILOT/ Born in Shawnee, Oklahoma, March 6, 1937. A USAF Lt. Colonel, he received a BS degree in Aeronautical Engineering from the Air Force Institute of Technology in 1956. Before joining NASA he flew F-84 and F-86 jets for four years with the 86th Fighter Bomber Group in Munich, Germany. He also served two years as an aeronautical engineer and test pilot for the Experimental Flight Test School at Edwards Air Force Base, California. His hobbies are as active as his work; he enjoys racing boats and cars as well as flying. Altogether he has logged flight time in excess of 5000 hours, 3000 in jet aircraft. Prior NASA work: Cooper piloted his *Faith 7* Mercury spacecraft through 22 orbits in 1963.

CHARLES CONRAD, JR.

PILOT/ Born June 2, 1930, in Philadelphia, Pennsylvania, he earned a BS degree in Aeronautical Engineering from Princeton University in 1953. Conrad entered the Navy following graduation and became a naval aviator. Presently, he holds the rank of Lt. Commander, USN. Before joining NASA he served as flight instructor and performance engineer at the Navy Test Pilot School at Patuxent River, Maryland, where he was also a project test pilot in the armaments test division. His flight time is in excess of 6000 hours, with more than 4800 hours in jet aircraft. Prior NASA experience includes monitoring the Apollo Command/Service and Lunar Excursion Modules. This will be his first space flight.

GEMINI 5
COOPER
CONRAD

GEMINI 5
Preparation

Previous technological accomplishments notwithstanding, no one underestimates the hazards of pioneering through the space frontier. Its sheer complexity magnifies the high drama: since any unanticipated problem is a potentially lethal situation, practice and more practice is required for all phases of the mission. Thus (**1**) Conrad and Cooper practice emergency water egress procedures assisted by two Navy frogmen, for one can drown in the sea upon returning from the oceans of space. The same months-long testing procedure is applied to machines as well the men. Modifications to Gemini launch vehicle (GLV) 5 are completed and stage I erected in the vertical test facility at Martin-Baltimore on January 5. Stage II is erected February 8. Power is applied to the vehicle for the first time on February 15, and subsystems testing is completed March 8. Another modification period follows. The Simulated Flight Test of Gemini spacecraft No. 5 begins at McDonnell on April 26, and is completed May 19; prepara-

1

2

3

tions for altitude chamber testing last until May 25. GLV 5 and spacecraft No. 5 are mechanically joined together (mated) at Kennedy Space Center's complex 19 on July 7 (**2**). Then Gemini-Titan 5 is taken apart from the spacecraft (demated) on July 23 to allow for modifications (**3**) and remated again on August 5. Meanwhile teams of ground support engineers and inspectors from the contracting firms continue the process of developing, refining, and practicing the checkout system of the combined Gemini-Titan unit. Of prime importance in the long path from manned orbital flight to lunar exploration is the development of simpler and more efficient checkout techniques so as to eliminate the excruciating chore of tearing through one perfectly good system to get at a defective part of another, a situation which plagued earlier Mercury shots. Gemini 5 is originally scheduled for launch on August 19, but despite all the pre-launch attention focused on it, a computer malfunction on that date causes a hold in the countdown. Coupled with deteriorating weather conditions on the Cape, this necessitates rescheduling of the mission to August 21. Seemingly oblivious to all this hassle, Conrad is fitted into his suit (**4**) while Cooper—if not so jovial as his partner—at least seems resigned to the long, multistaged routine of fitting and checking his space gear and himself (**5**). At Kennedy, they're faster with a thermometer than the proverbial school nurse. Meanwhile, (**6**) the carefree crowd awaits, anxious for the show.

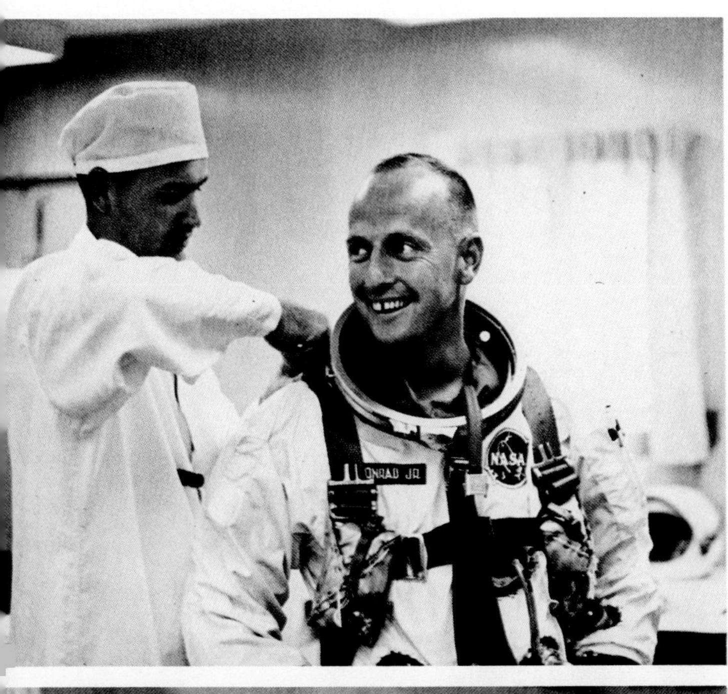

6

GEMINI 5
Pre-launch / Launch

The moment arrives. Striding purposefully up the walkway with their air conditioning units in hand, the two almost seem like executives toting briefcases, about to catch the morning commuter train. But the support crews around them snapping pictures (**1**) testify this is no ordinary departure. Though their yellow shoes moving along the sunrise orange path are rather impish, the event's seriousness is reflected in the faces of personnel at launch control where men hunch over their monitors (**2**) keeping careful watch of time and the weather. Settling into the capsule (**3**), talking among themselves and launch control, the two astronauts begin the intricate procedures of pre-launch. Even the great Titan rocket seems dwarfed now by the structures surrounding it at the launch site (**5**). But then, at 9 AM EST, the rumble of ignition is heard across the Cape—and it is the center of attention, spawning a pillar of light and hurtling into the clear Florida sky (**4**), its signature a squiggle against the immensity of space.

1

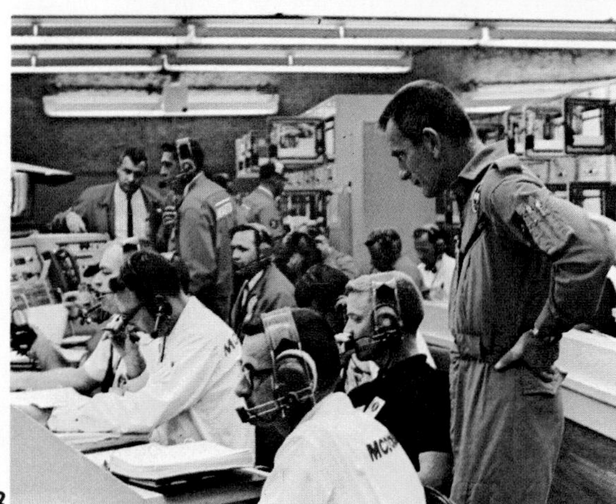
2

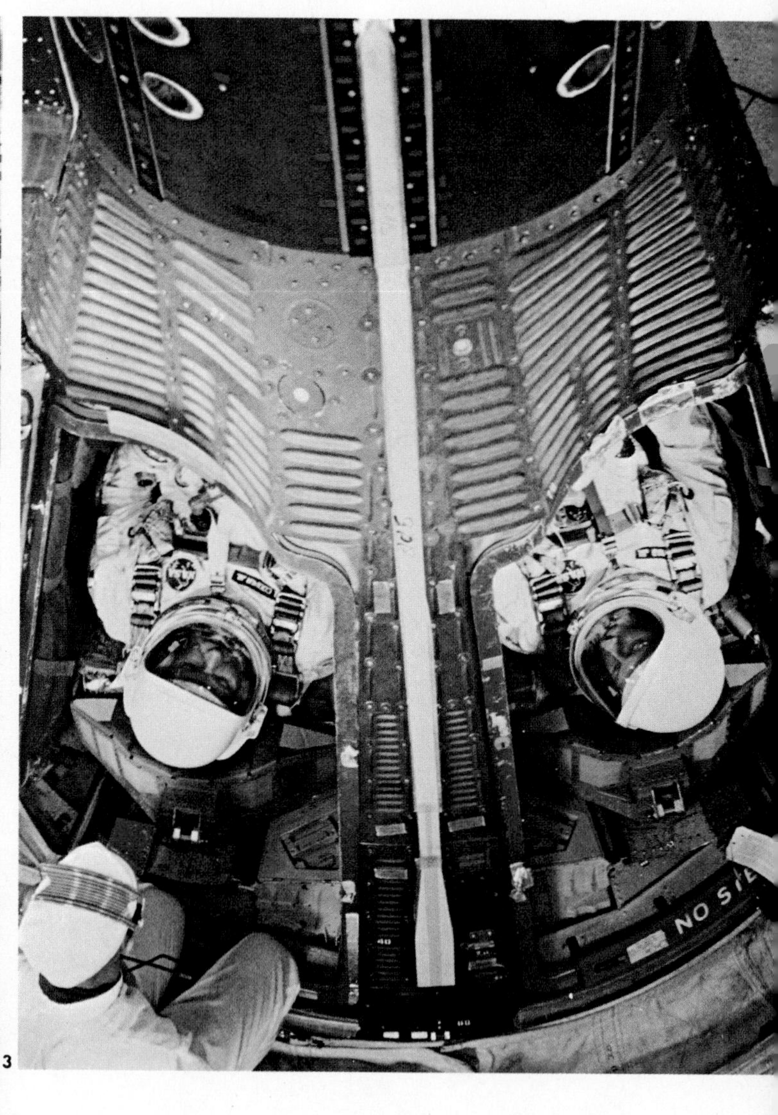
3

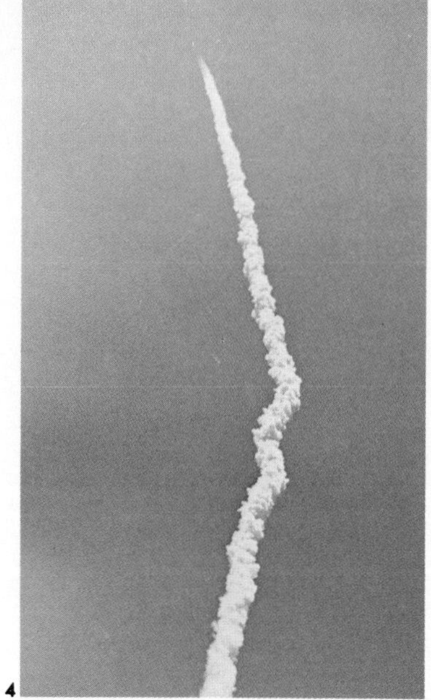
4

GEMINI 5
Mission

While Conrad and Cooper are achieving orbital velocity, the mission takes an ususual turn at earth's surface. Previously, those parts of the booster rocket which had served their purpose just fell away into the sea and sank. However, with Gemini 5 is achieved the first successful recovery of a spent second stage (**1**), looking much like a lead balloon. But there's nothing leaden about Gemini's swift flight as it speeds above clouds over the coast of southern California. In a view from Malibu to Mexico, Santa Catalina and San Clemente islands lie off Los Angeles' Palos Verdes peninsula (**2**), the jagged blue crescent of San Diego Bay marks the coast right of center, and the Salton Sea in Imperial Valley lies beyond. A modified Hasselblad 70 camera was used with MS Ektachrome thin base film. Astronaut Charles Conrad Jr., his face illumined like a Rembrandt portrait (**3**), turns momentarily from his observations; perhaps he's reflecting on the difference a day makes as the spacecraft sails over Florida

1

2

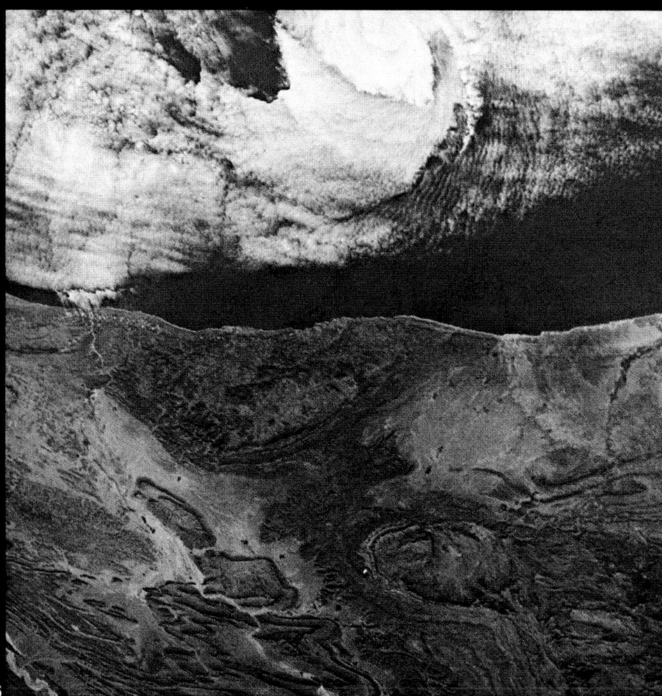

5

6

and Cape Kennedy (**4**). The ability to photograph vast areas of the globe from space, with an accuracy denied land-based cameramen, is one of the program's most exciting accomplishments. It frees cartographers from centuries of terrestrial restrictions, aids geoscientists exploring little-known regions, and is especially useful to meterologists now able to view cloud formations hundreds of miles across. A prime example is this stratocumulus cloud spiral (**5**) off the coast of Morocco, representing a lee eddy induced by the airflow around Cape Rhir to the north. Among Gemini 5's main objectives is the testing of a radar and guidance system developed for rendezvous and docking with an orbiting Agena rocket. To that end it carries a rendezvous evaluation pod (REP) in the adaptor section of the craft, containing a radio, transponder, batteries, antenna, and flashing lights—seen here in an artist's representation (**6**). Though the REP is successfully ejected, failing pressure in the oxygen supply tank of the fuel cell system causes abandonment of the exercise. However, in the third day of the mission a simulated rendezvous with an imaginary Agena is conducted at full power using the orbit attitude and maneuver system (OAMS). Activities through day four concern mainly operations and experiments; on day five OAMS operation becomes sluggish and thruster No. 7 inoperative—followed by loss of thruster No. 8 the next day, as the system gradually becomes more erratic. Mission Control (**7**) orders retrofire during the eighth day.

6

4

GEMINI 5
Splashdown

On August 29 the spacecraft burns through the morning sky, splashdown occuring at 7:56 AM EST in the western Atlantic, some 91 miles short of the projected landing point. Recovery swimmers from the aircraft carrier *Lake Champlain* secure a floatation collar to Gemini 5, seen here bobbing in the helicopter's downdraft as a member of the recovery team leaps to join operations for the last portion of the flight (**1**). Should surface waves be violent, the diving gear enables pararescuemen to attach the collar working underwater; the Air Rescue Service has worldwide recovery commitment for all Gemini-Titan flights. Altogether it requires 90 minutes to fly the astronauts from spacecraft to ship; at 9:25 Cooper and Conrad step into morning sunlight on the deck of the *Lake Champlain* to the traditional red carpet welcome (**2**). In another 90 minutes, at 11:51, their craft too will be lifted aboard. It is a time of gladness and camaraderie, and a little clowning. In just a simple gesture such as Conrad's playful tug at his partner's

1

2

5

eight-day growth of beard (**3**), the immensity of this longest flight yet is rendered in human scale. But a spaceflight doesn't end with a successful recovery and a flurry of cheers; there is still work to be done in the post-flight operations. The complex human machine must be checked for effects of of prolonged confinement, dehydration, fatigue and weightlessness; there are comparisons of pre-flight and post-flight blood pressures, blood volumes, pulse rates, and electrocardiograms, as there are with every mission. Information from tests of bone demineralization, work tolerance, and heart muscle deterioration (with the aid of a phonocardiogram) must be assimilated. New to this flight is cardiovascular conditioning, in which pneumatic cuffs were applied to upper thighs and automatically pressurized to 80mm Hg for two out of every six minutes of the awake cycle, to determine their effectiveness in preventing deterioration induced by prolonged weightlessness. Another new experiment concerned changes in otolith (calcerous gravity gradient sensors in the inner ear) functions. In this the astronaut attempted to judge the spacecraft's correct pitch while wearing special blindfold goggles. The follow-up to these experiments is initiated aboard ship (**4**). Afterwards the celebration continues with a congratulatory call from President Johnson (**5**). Reunited with their wives, and joined by Cooper's daughters, they prepare to board Air Force One (**6**). It will carry them from Houston to Washington for another round of acclaim.

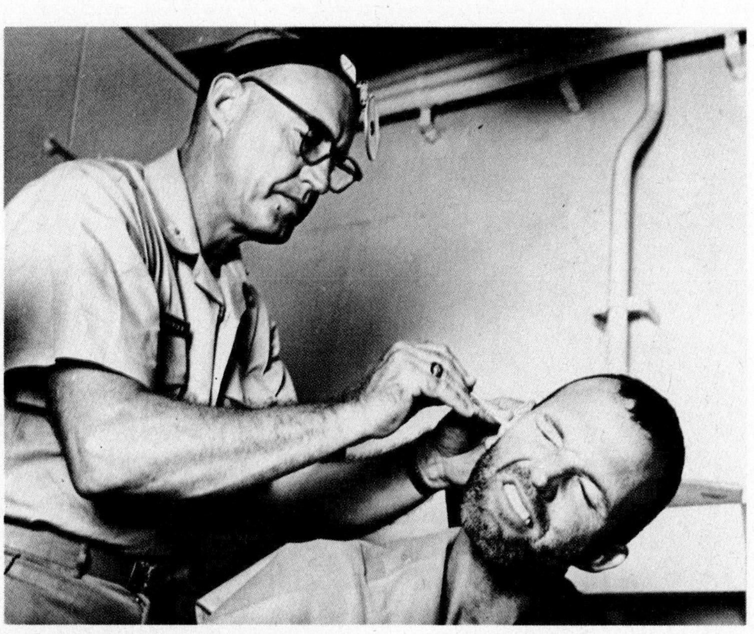

4

GEMINI 6

After three developmental flights and two marathon missions to collect scientific, medical and military data, the Gemini program moves into its most ambitious undertaking with the scheduled flight of Gemini 6. Its two-day mission is to begin the development of a series of rendezvous and docking exercises in which the Gemini space capsule will overtake, join and lock onto a previously launched Agena target vehicle modified for this purpose. Perfecting the maneuver is vital if America is to reach the moon by 1970, as a direct flight of that distance would require a booster with a 15-million-pound thrust, and no such hardware can be developed in time. To meet the stated goal on schedule, a technique called "lunar orbit rendezvous" will allow the Saturn V booster already under development to place a capsule into orbit around the moon. From this, a separate vehicle can leave and return using the rendezvous and docking maneuver, showing its importance as a prime objective. Four scientific experiments are also scheduled, but extravehicular activity (EVA) is not planned until Gemini 8. While Ed White has taken a 21-minute space walk during Gemini 4, his trip outside the spacecraft resulted from a last-minute NASA decision to recapture the initiative lost when Russian cosmonaut Colonel Alexis Leonov left his Voskhod 2 for 12 minutes during its second orbit, some five days *before* Gemini 3 took to the sky. But further American space spectaculars will have to wait their turn. Command pilot for Gemini 6 is Wally Schirra, who flew the fifth Mercury flight in *Sigma 7;* making his first space trip, Tom Stafford will accompany Schirra as pilot.

WALTER M. SCHIRRA, JR.

COMMAND PILOT/ Now approaching his 43rd birthday, Navy Captain Schirra is one of the seven Mercury astronauts selected in 1959. A native of Hackensack, New Jersey, Wally was graduated from the United States Naval Academy in 1945 with a Bachelor of Science degree and entered flight training at Pensacola. After 90 combat missions over Korea, for which he received the Distinguished Flying Cross, Schirra returned stateside, serving various tours as a test pilot while logging over 3800 hours in the air. Wally flew the 6-orbit *Sigma 7* mission on October 3, 1962 and served as the backup command pilot for Gus Grissom on Gemini 3. He and his wife Josephine Schirra have two children.

THOMAS P. STAFFORD

PILOT/ USAF Major Thomas P. Stafford of Weathersford, Oklahoma, was commissioned in the U.S. Air Force after graduation from the United States Naval Academy with a Bachelor of Science degree in 1951. He flew with fighter interceptor squadrons in the U.S. and Germany, attended the USAF Experimental Flight Test School at Edwards AFB, California, and joined the staff of the USAF Aerospace Research Pilot School at Edwards. As Chief of the Performance Branch, he supervised and administered the curriculum. Co-author of the *Pilot's Handbook for Performance Flight Testing* and the *Aerodynamic Handbook for Performance Flight Testing,* Tom joined NASA in 1962, and is the father of two children.

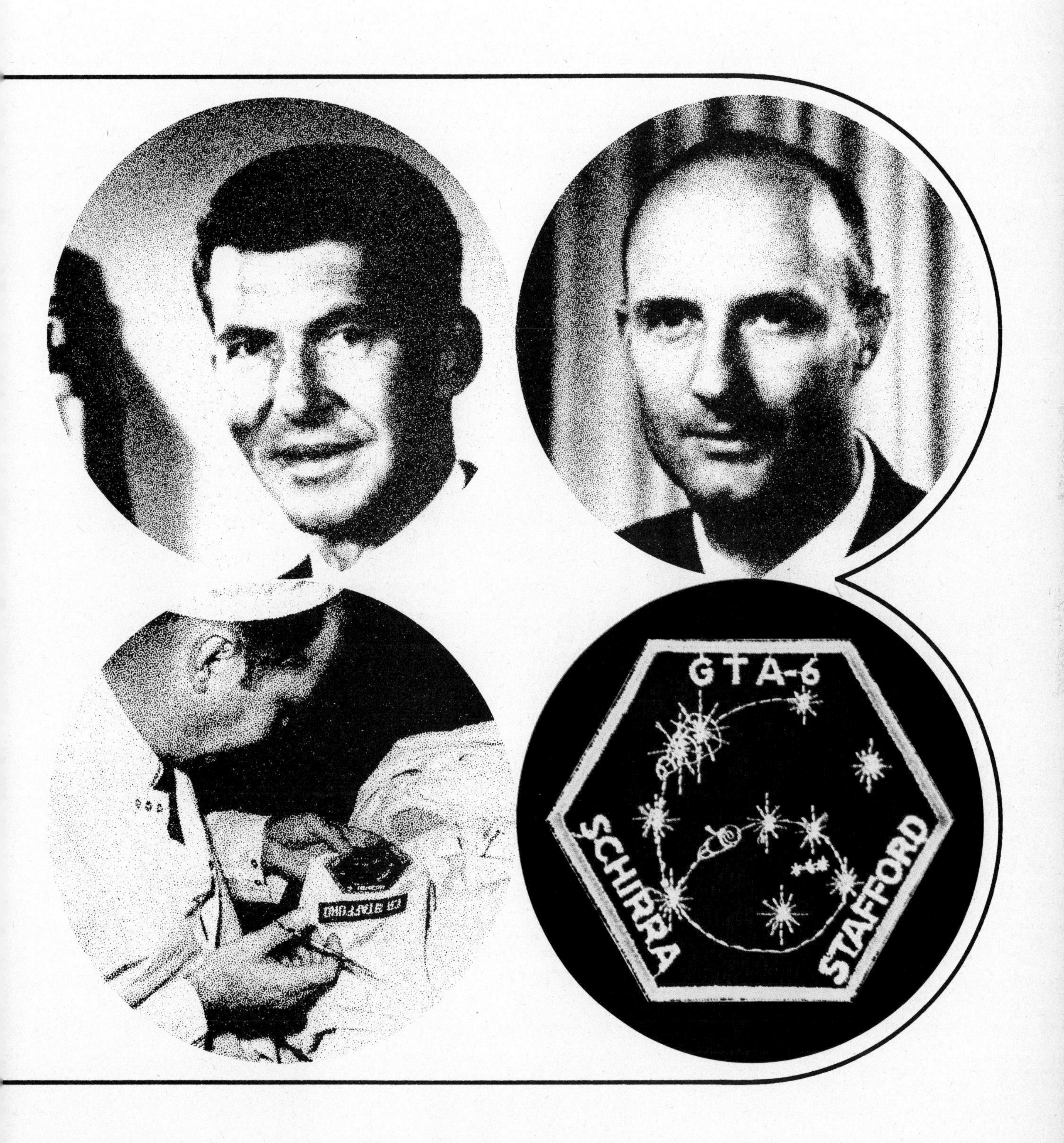
GTA-6
SCHIRRA
STAFFORD

GEMINI 6
Pre-launch

On July 26, 1965, Gemini Agena Target Vehicle 5002 is designated as GATV for Gemini 6; Atlas 5301 will serve as its launch vehicle. The Agena has been modified with a docking collar to latch onto the nose of Gemini for rendezvous. Stage I and II of Gemini launch vehicle (GLV) 6 is erected at Complex 19 on August 30-31 but Hurricane Betsy sweeps near the Cape, delaying installation of the spacecraft atop GLV for 7 days (**1**); this is not accomplished until September 9. As launch date nears (**2**), Schirra (at center) and Stafford (right) complete their training but remain on a limited food intake in preparation for their upcoming flight. Each man is keenly aware of the importance this mission holds, but their tension does not show. Lift-off is scheduled for 11:35 AM EST on October 25. Wally and Tom arise early that day, and five hours before launch, the countdown begins when the PA system announces, "T minus 300 minutes and counting." Suiting up, the two astronauts arrive at the launch deck by van

1

2

3

and are transferred by elevator to the "white room" atop the GLV. At T-110 minutes (**3**), the capsule is ready for Schirra (left) and Stafford (right) to enter. Technicians remove Wally's boot covers and he steps inside, followed a few moments later by Tom Stafford. The two men carry out their pre-flight checks and the countdown proceeds smoothly. As the GATV must complete one orbit before Gemini races after it, the Agena is to be launched 95 minutes before Gemini 6 lifts off. When the Atlas-Agena countdown reaches zero (**4**), its three main engines are ignited. Ice from the liquid oxygen tanks covers the upper half of the Atlas, and vapor streams from its vent valves. The shackles holding it in place are released and moments later, at 10:00 AM EST, the rocket thunders into the sky. Five minutes after, the Agena separates on schedule from its Atlas booster. Ground control (**5**) is in constant contact and all signals appear normal. The air of anticipation that accompanies every launch turns into a quiet sigh of relief, but 75 seconds later, telemetry contact with Agena is lost and cannot be regained. Suddenly, all trace of the Agena disappears from the Atlantic tracking network's radar screens. The Gemini 6 countdown is held while a hurried conference between Houston and the Cape considers the possibility of substituting another craft as target vehicle. But at 10:54 AM, the flight is scrubbed and the two astronauts disembark. Something unexplained has gone wrong with the Agena—and no trace of it will ever be found.

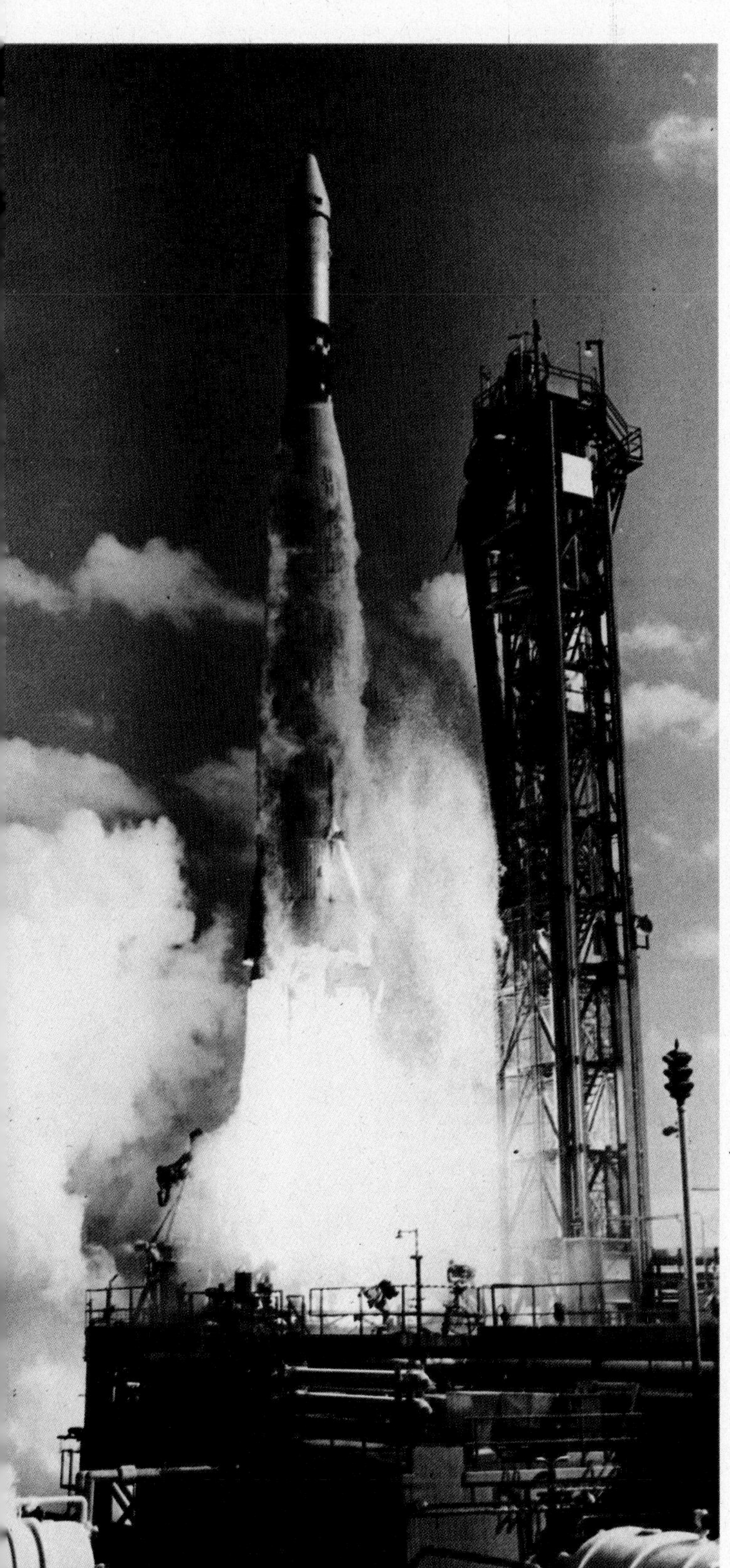

5

GEMINI 7

The unexpected failure of the Agena to orbit, halting the flight of Gemini 6, is viewed by NASA as a major setback. One more mission, Gemini 7, is scheduled to lift off December 4, 1965. A long-duration flight, this is to serve as the final test of man's adaptability to the environment of space. Extensive bio-medical tests are to be run during the 14 days in space and 20 experiments are assigned, 14 of which have already been performed on earlier missions. These continuing experiments are designed to check data under differing flight conditions and uncover variable factors. The new lightweight (16-pound) removable pressure suits under development are now ready for use and will be worn for the first time by command pilot Frank Borman and pilot James Lovell, Jr. As a space rendezvous is desperately wanted before the end of the year, NASA officials immediately begin discussing ways to turn the defeat into triumph after the October 25 failure of Gemini 6. Drawing on earlier preliminary planning for a rapid manned flight launch, NASA Administrator James Webb dictates a memorandum to President Johnson in which he suggests the possibility that Gemini 6 can be rescheduled to launch in time for a rendezvous of two manned spacecraft—if the lift-off of Gemini 7 does not cause too much damage to the launch pad. On October 28, the White House announces that Gemini 6 will be launched while Gemini 7 is in orbit and will attempt a rendezvous during which the Gemini 7 crew will continue with their assigned duties, maneuvering only to make their capsule a better target. This is the big test for rendezvous. Everyone wonders, "Can it be done?"

FRANK BORMAN

COMMAND PILOT/ USAF Major Frank Borman was born March 14, 1928 in Gary, Indiana. After graduation from the United States Military Academy in 1950 with a Bachelor of Science degree, Borman completed pilot training at Williams AFB, California. Flying fighter craft until 1956, Frank earned a Master of Science degree in Aeronautical Engineering at California Institute of Technology in 1957 and became a professor at West Point. Then a test pilot, he was selected by NASA in September 1962, and now has over 4400 flight hours. The father of two children, Frank will wear the Gemini 7 emblem—a torch and capsule set in the background of space. He was the backup commander of Gemini 4.

JAMES A. LOVELL, JR.

PILOT/ A Navy Lieutenant Commander, Lovell is a native of Cleveland, Ohio, and will be 38 next March 25. A 1952 graduate of the United States Naval Academy (Bachelor of Science degree), Lovell completed flight training and has been a test pilot and program manager for F4H weapon system evaluation. Jim also served as a flight instructor and safety officer at the Naval Air Station, Oceana, Virginia, with over 3000 hours of flight time to his credit. Specializing in recovery problems for NASA, Lovell was selected as an astronaut at the same time as Frank Borman; this is the first space flight for both men. Jim has three children, plays handball and tennis. He was backup pilot for Gemini 4.

VII

GEMINI 7
Launch

Despite the now-scheduled rendezvous with Gemini 6, no changes are made in the flight plan of Gemini 7 (**1**). Astronauts Lovell (third from left) and Borman (sixth from left) continue their regular training and conference sessions while frenzied NASA activity behind the scenes prepares for the dual-launch mission. As complex 19 is the only pad capable of Gemini-Titan lift-offs, Gemini 7 must be sent into orbit and the pad repaired in record time to accommodate Gemini 6. Quick pad repairs are not the only problems facing the ground staff; after Gemini 6 is placed in firing position, a series of pre-flight checks numbering about 400,000 must be carried out. But confidence is high and preparations for the Gemini 7 launch proceed. GLV-7 is erected on October 29-30, and 12 days later (**2**), the Gemini spacecraft is hoisted atop its Titan booster, with connection and testing procedures beginning immediately. The rest of November is spent in testing and verification procedures, which are simplified because there is no simultane-

1

2

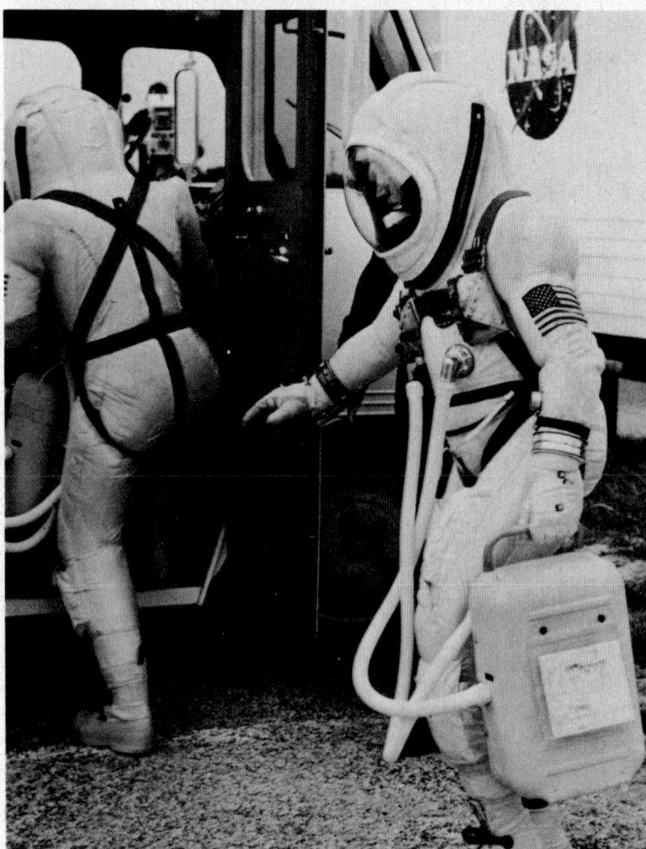

3

4

ous Atlas-Agena launch. These are completed on the 27th while the ground staff makes plans to compress all this activity into a 10-day span for the Gemini 6 launch to follow. A few hours before the scheduled 2:30 PM EST launch of Gemini 7, the two astronauts enter the suiting trailer at launch complex 16, where they don the new G5C space suits. Carrying their portable cooling systems (**3**), Frank Borman (leading) and Jim Lovell enter the transport van which will carry them to the ramp that leads to the elevator at Complex 19. A short trip up the erector and several checks later, both men are secured in their Gemini capsule awaiting the end of the countdown and the start of their 14-day flight. At 38 minutes before blast-off, the launch area is cleared and the drama shifts to the blockhouse (**4**), where Fritz Widick (NASA Test Conductor), Roy Post (McDonnell Systems Test Engineer) and astronaut Alan Bean (left to right, foreground) check data transmitted from the Central Control building located a mile away. The countdown continues while tracking stations report, radar is checked, weather conditions are monitored and hundreds of lights blink a reassuring green to indicate that everything is working correctly. As the countdown ends, an electrical signal to the Titan opens its Stage I engine valves. Propellants flow into the thrust chambers and ignition takes place. Flames shoot up from the Titan as its engines build up thrust and Gemini 7 blasts off (**5**), with the pad abort rescue teams watching from their posts.

GEMINI 6
Mission

As the Gemini 7 spacecraft roars high into the sky atop its Titan booster, the diminishing fuel supply of the Stage I engines lighten the rocket and the Titan continues to accelerate until it reaches a point approximately 50 miles high and 50 miles downrange from the Cape. Traveling at about 6000 mph, it suddenly runs out of fuel and the Stage I engines shut down. Inside Gemini, the two astronauts experience a momentary relief from the tremendous G forces. Explosive bolts holding the two stages together are triggered at the same time as fuel valves to the Stage II engine open. Combustion in the Stage II sustainer engine occurs instantly, as does a thrust build-up to 100,000 pounds and G forces return inside the capsule. Stage I engines and tanks drop away and Stage II, with its Gemini space capsule attached, races along to its destination some 87 miles above the earth. A radio signal from Ground Control shuts down the booster engine and 30 seconds later, Borman and Lovell separate from the second stage.

1

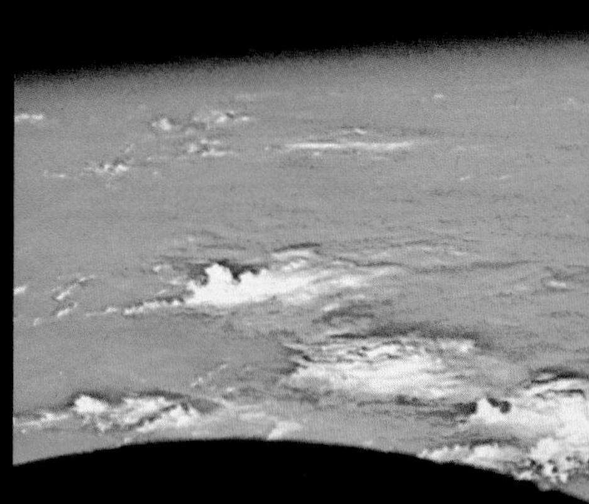

2

immediately after separation, Frank maneuvers the capsule around 180 degrees so that its blunt end faces forward. He then approaches within 60 feet of the booster second stage to begin "station keeping," as formation flying one of the secondary mission objectives. Fifteen minutes later, Gemini is powered down to prepare for its 14-day mission and enters its first darkness period. Confined to a cabin space only slightly larger than a telephone booth and speeding along their orbit at almost 18,000 mph, Borman and Lovell set about conducting the experiments that are to occupy them for the next nine days while they await the arrival of Gemini 6. Although both men have requested permission to remove their space suits, Gemini Program Office is afraid that any possible damage or leakage from the pressurized spacecraft may incapacitate them quickly, and so the mission plan requires one man to be suited at all times. About 45 minutes Ground Elapsed Time (GET) into the flight, Jim Lovell removes his pressure suit. He will work without it for about 100 hours then suit up while Frank removes his. As Gemini 7 passes over the east coast of Florida (**1**), terrain photography captures the coastline from St. Augustine to Fort Pierce. Then, in one of the most spectacular views (**2**) they have ever seen, the two astronauts photograph the earth and moon together. Off the coast of California the island of San Clemente is seen (**3**). As Gemini 7 passes over Kennedy Space Center (**4**) on December 12, they film the abort of Gemini 6

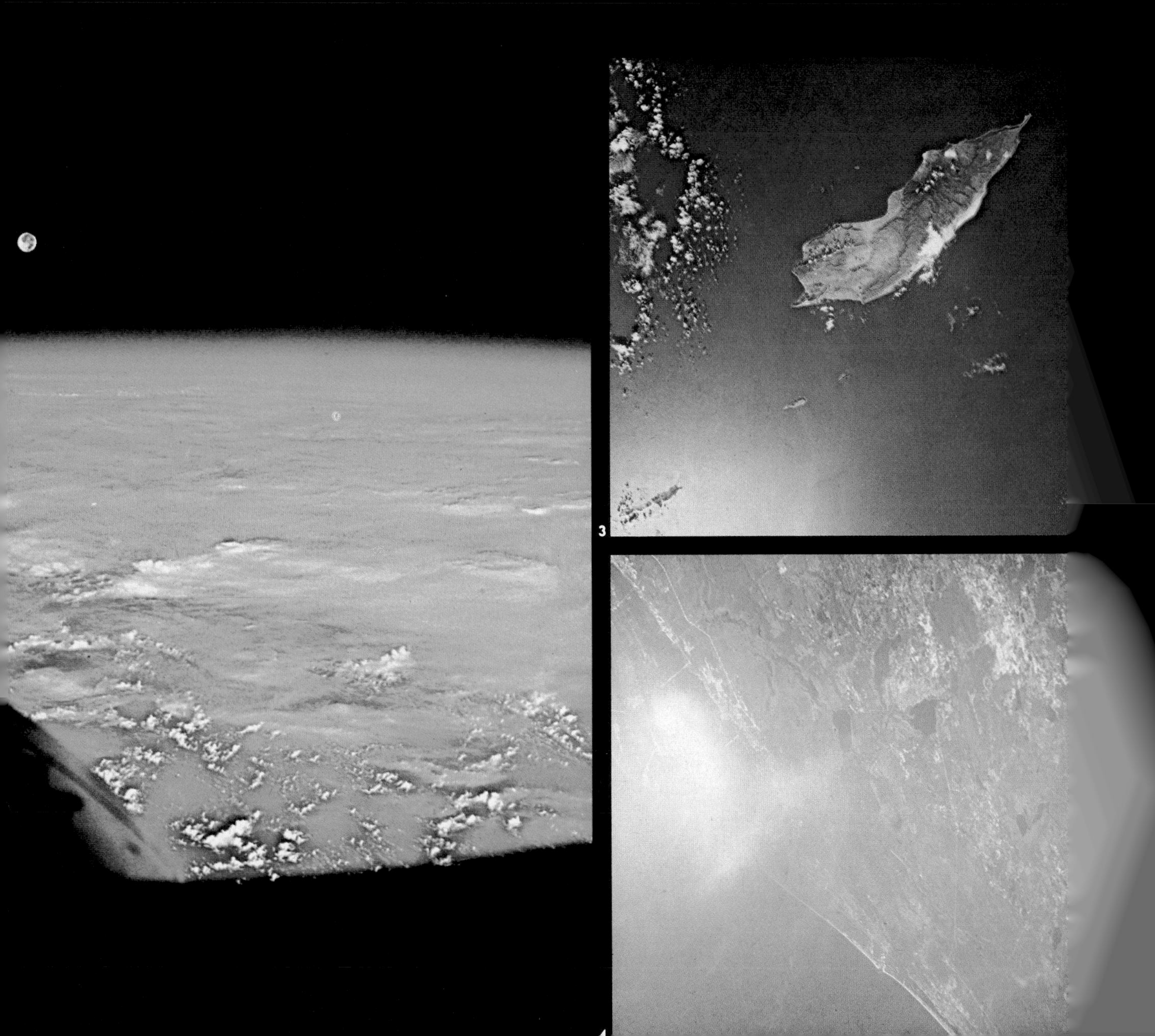
3

GEMINI 6
Launch

The launch pad at Complex 19 is still hot from Gemini 7's blast-off when ground crews begin repairs that will allow the launch of Gemini 6 on December 13, 1965. Within 24 hours, both stages of GLV 6 are erected for the second time in 42 days, the Gemini capsule is connected and power applied. Cutting corners wherever possible in the pre-launch check routine, subsystems testing is completed December 8. Discovering a malfunction in the Gemini computer memory, the entire unit is replaced. With the end of the Simulated Flight Test on December 9, all pre-launch tests are completed and lift-off is moved ahead one day to December 12. Again (**1**), Wally Schirra (leading) and Tom Stafford walk up the ramp for the elevator ride to their space capsule and for the second time, Schirra and Stafford ease into the couches inside the spacecraft, connect their suits to support/communication cables and await the end of the countdown. As NASA has learned that malfunctions with the Titan propulsion system hap-

1
2
3
4

pen just after engine ignition, the GLV is always tethered to the ground for a few seconds' "hold-down" after ignition to let the blockhouse monitor critical points. At 9:54:06 AM EST, the "fire" signal is given by the countdown computer, and the Titan's engines blaze to life. One second later (**2**), the Malfunction Detection System shuts the engines down; an electrical umbilical connector has dropped off prematurely. Schirra now has a hard decision to make fast—to pull the ring between his knees, ejecting himself and Stafford from the capsule, or to stay in place and pray that no explosion occurs. Knowing that ejection will mean no flight until early 1966, and determined to rendezvous with Gemini 7 (now passing overhead) if possible, he elects to stay. Thankfully, the fully-fueled Titan behaves itself. Replacing the umbilical cable is accomplished quickly, but routine data analysis reveals a lack of thrust in one engine. Dismayed but not discouraged, Wally and Tom are removed from their spacecraft—again—and the launch is rescheduled for December 15. A protective dust cap left on an inlet port is the cause of the thrust problem. Three days later, Wally and Tom again await the blast-off of Gemini 6. At 8:37 AM EST Gemini 6 leaves (**3**) in what turns out to be a routine launch, but those in Launch Control watching this third attempt (**4**) monitor every facet. Heading toward orbital altitude (**5**), Gemini 6 and its chase planes produce a dramatic contrail over Cape Kennedy. Third time lucky, Wally and Tom aim at rendezvous.

GEMINI 6/7 Mission

Now designated as Gemini 6A, the spacecraft carrying Wally Schirra and Tom Stafford joins Gemini 7 in orbit at a point some 1200 miles behind Borman and Lovell. While Tom operates the onboard computer, Wally maneuvers the capsule, using its thrust rockets to adjust the orbit and bring Gemini 6A into position for rendezvous. Three hours into the mission, the onboard radar is turned on and with 246 miles separating the two spacecraft, radar lock-on is achieved. Further maneuvering by Schirra is followed by final braking, and at 5:50 GET, rendezvous is technically accomplished. Borman greets his fellow astronauts over the intercom with, "What kept you?" Formation flying begins at 5:56 GET with the two Gemini capsules only 120 feet apart (**1**), and Wally photographs the Gemini 7 craft through his hatch window. The two Gemini capsules are now traveling at an altitude of approximately 160 miles above the earth and at a speed of almost 18,000 miles per hour. The orbital maneuvers controlled by computer are more

2

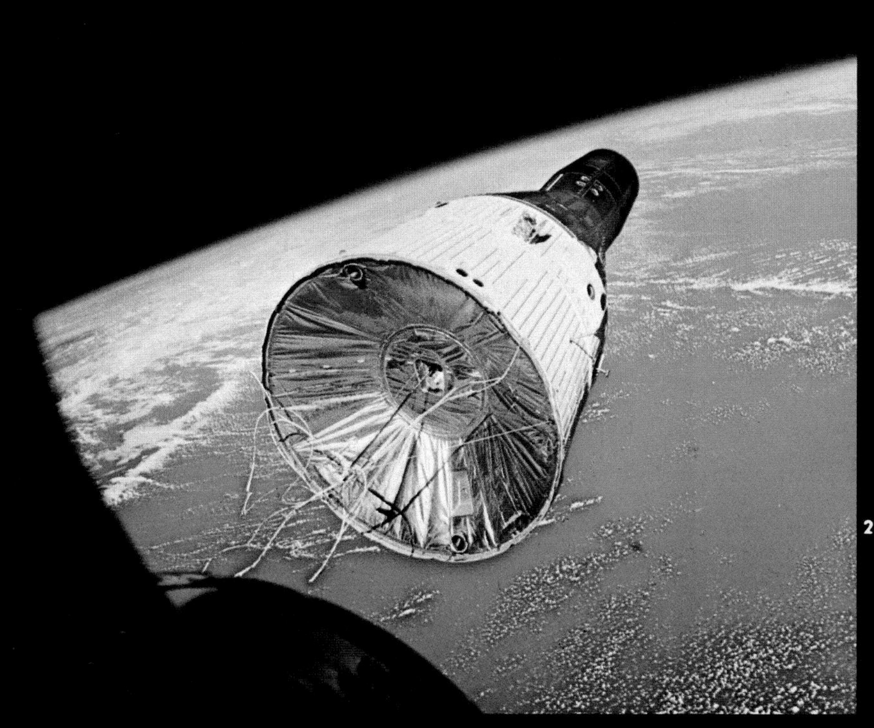

5

precise than anticipated and in a series of slowly-executed half-circles, Gemini 6A pulls ahead (**2**) of its space partner. Still operating the onboard computer, Stafford lets the world know that he is "busier than a one-armed paper-hanger" but the success of rendezvous maneuvers to this point confirms the confidence of all four men involved. Passing above Gemini 7 (**3**), Schirra photographs the target capsule against a cloud mass between it and the land formations below. Continuing its maneuvering with extreme accuracy (**4**), Gemini 6A pulls alongside the target capsule. The crew of each spacecraft can now see one another and they wave greetings through the hatch. Wally repeatedly approaches to within a few feet of the Gemini 7 capsule, coming as close as 10 inches at one point, but no attempt at docking is made. NASA fears that should the two spacecraft actually touch, a static electricity discharge may occur with unknown consequences, and so mission plans do not include docking on this flight. For their final rendezvous orbit, Gemini 7 now flies formation on Gemini 6A and as Borman and Lovell pull ahead (**5**), Schirra and Stafford insert a "Beat Army" sign in their hatch window, adding a touch of interservice humor. With the mission completed, the Gemini 6A crew waves good-by and changes into a higher orbit to prepare for re-entry. Ever the gagster, Schirra calmly reports the sighting of a UFO powered by a team of reindeer, causing a momentary panic at Mission Control until they catch on.

3

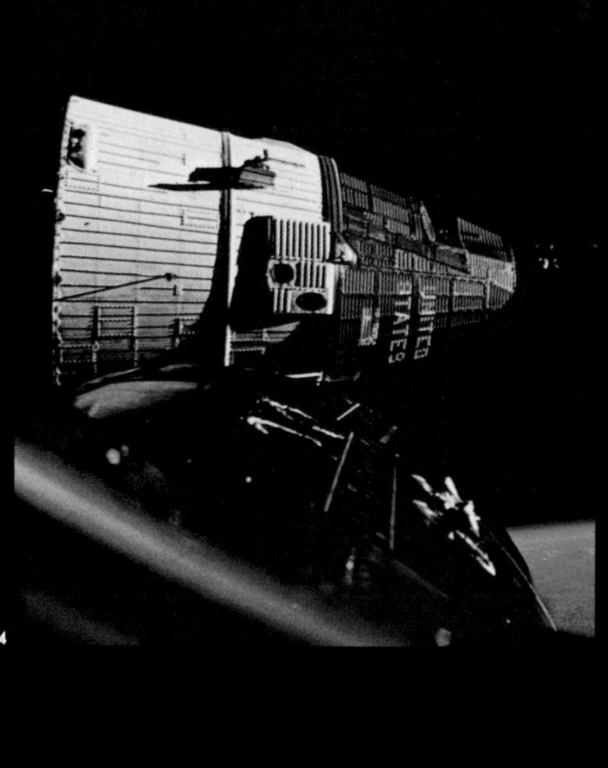
4

GEMINI 6 Splashdown

With only one exception, the flight of Gemini 6A is an unqualified success. At 20:55 GET into the mission, its delayed-time telemetry tape recorder stops working and for the remaining 4 hours, 20 minutes in space, all telemetry data is lost. But Schirra and Stafford are elated, both men having anticipated problems that did not occur in accomplishing the rendezvous. They were especially concerned about the amount of fuel that would be required to maneuver the capsule while flying formation. Now the mission is over, and with 3½ successful rendezvous orbits behind them, Wally finds that one-half of the allocated fuel supply remains unused. After playing two choruses of "Jingle Bells" on a harmonica brought along for the occasion, the two men prepare for splashdown in the Atlantic. Schirra maneuvers the capsule around to face its blunt end forward for the plunge into the earth's atmosphere. This is crucial, as the blunt end carries the heat shield, and should the spacecraft tumble out of control during reentry, it could burn up.

1

2

As Gemini 6A slows down and begins to fall toward earth, Wally jettisons the adapter section and at 10:29 AM EST on December 16, Gemini 6A hits the ocean water seven miles from its planned landing point, ending a 17-orbit, 449,800 mile flight that lasted 25 hours, 52 minutes. A three-man U.S. Navy frogman team (**1**) is first on the scene to help in the recovery operation and the two astronauts open their hatches while awaiting the arrival of the USS *Wasp,* prime recovery ship. Less than an hour after splashdown, the *Wasp* arrives on the scene (**2**) as its recovery helicopter descends to pick up Schirra and Stafford. While the helicopter returns to the *Wasp* with the two astronauts, their Gemini re-entry capsule is recovered (**3**) and hoisted aboard ship for return to NASA and a thorough post-flight evaluation. Safely aboard the carrier (**4**), Wally Schirra (leading) and Tom Stafford make their way through the assembled crew and reporters for the official welcome-home ceremony. While everyone concerned is jubilant about their successful mission, no one is more pleased than Schirra, whose decision not to eject from the capsule during the aborted launch of December 12 made this occasion possible. Standing behind a hastily-erected podium (**5**) and a gigantic cake prepared in their honor by the ship's cooks, Schirra (left) and Stafford (right) convey their appreciation for the smooth and quick recovery from the ocean. From here, it's back to the Cape and a reunion with their families just in time for enjoyment of the Christmas holidays.

5

GEMINI 7
Splashdown

Waving good-by to the Gemini 6A spacecraft as it moves out of formation in preparation for re-entry, Borman and Lovell complete the most intensive program of medical space research yet undertaken. As their 14-day mission draws to a close, the two astronauts have successfully overcome two major problems of concern to medical experts: the effect of weightlessness on the human heart, and the effect a lack of exercise has on the kidneys. Previous flights have indicated that long missions may cause extensive heart strain and bone calcium dissolution leading to kidney stones. But Borman and Lovell disprove these fears. Their one major problem seems to be Frank's insomnia. As Borman points out, "People can project the problem if they imagine living two weeks in the front seat of a Volkswagen. . . . I don't know why I had difficulty in sleeping. Jim Lovell slept very well . . . . Perhaps the lack of space and the fact that we weren't able to stretch out at all led to it . . . ." As Gemini 7 nears the end of its

1

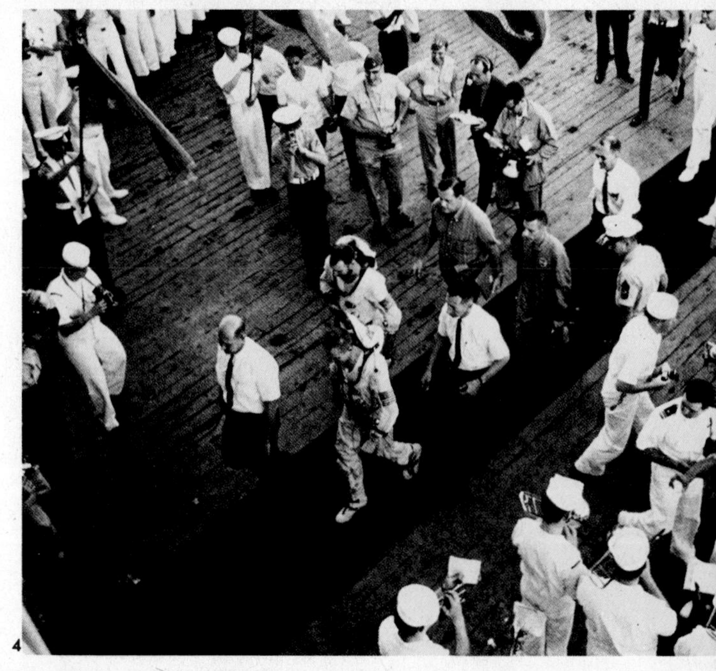
4

206th revolution around the earth, Frank Borman pushes the retrofire rocket button to begin their re-entry maneuver, and at 9:05 AM EST on December 18, the two astronauts splash down in the ocean 700 miles southwest of Bermuda, only seven miles off target. As with Gemini 6A, the Navy frogman team arrives on the scene first (**1**), helping Frank (right) and Jim (left) into a life raft to await helicopter pickup and return to the USS *Wasp*. A flotation collar attached to the capsule increases its buoyancy prior to recovery. When the Navy helicopter arrives (**2**), Jim Lovell is hoisted aboard first, as Borman waits in the raft below for his turn (**3**). After landing on the recovery ship, the two men disembark from the helicopter and with Frank striding slightly ahead of Jim (**4**), they take the same red carpet walk on the carrier deck taken by the Gemini 6A crew two days before. Bewhiskered but happy (**5**), the astronauts pose briefly for the cameras, as NASA public relations officer Ben James stands behind, and Dr. Howard Minners (right) watches their reactions with interest. A full-dress welcome-home ceremony awaits them. Meanwhile, the *Wasp* has pulled alongside their spacecraft and after the frogmen secure the hoist line (**6**), it is brought aboard and lowered carefully onto a dolly for return to NASA and an extensive post-flight evaluation. After logging 206 orbits and 5,716,900 miles, Gemini 7 is finally over, and its crew shares a monumental new record for time in space—330 hours, 35 minutes and 17 seconds.

3

6

THE AGENA TARGET VEHICLE

The Gemini helpmate that sometimes doesn't

One of the most important phases of the Gemini Program is the docking phase (**1**). Project Apollo will require two separate spacecraft to be docked or "coupled" while in space. Development of in-space navigation, maneuverability and the necessary hardware to achieve safe and positive coupling of two vehicles is imperative if we wish to go to the moon. The Gemini capsule is equipped with a number of small rocket motors called "thrusters" which give it very sensitive maneuverability. Developed concurrently for docking maneuvers is the Gemini Agena Target Vehicle (GATV). This unmanned vehicle is launched by a suitably modified Atlas ICBM, and placed into a waiting orbit for the Gemini manned capsule. It is a modified USAF Agena D upper stage, powered by a restartable Bell liquid-fuel engine (**2**) which delivers 16,000 pounds of thrust. The Agena's overall length with docking equipment is 36.3 feet. Its weight in orbital configuration is 7000 pounds. Along with its main powerplant, the Agena is equipped

1

with a pair of 200-pound-thrust aft-directed engines for small directional changes. For very fine or "vernier" adjustments, there are two tiny 16-pound-thrust engines. These are further used for "ullage" maneuvers. Ullage is a short jerking maneuver that settles the fuel and oxidizer in the tanks, ensuring proper main engine starting. Six nitrogen jets, coupled with the pivoting action of the main engine, allow roll, pitch and yaw maneuvers. The docking collar (**3**) with its notched alignment cut is designed to dock with the forward section of the Gemini spacecraft. Rendezvous, alignment and docking are accomplished with the aid of radar and powerful strobe lights on the Agena's hull. The Agena can be fired by either the Gemini crew or Mission Control. The Atlas Standard Launch Vehicle (SLV-3) is a modified ICBM (**4**). Included in the modifications are a special autopilot, increased structural strength in order to support the Agena, and relocation of the retrorockets to the upper interstage adaptor section. This will ensure separation of the two vehicles after the Atlas section has completed its launch burn. The Agena is then able to place itself into orbit under Mission Control direction, using its own main propulsion system. At launch, the docking collar is covered with an aerodynamic shroud, which is in two sections. After orbit is achieved, the shroud sections are jettisoned by explosive bolts and spring-loaded clamps. The Agena is under USAF management via their Space Systems, and is built by Lockheed's Missile Division.

3

GEMINI 8

Gemini 8 will be the sixth manned flight in the Gemini-Titan missions. By now, many of the intended objectives have been attained, and confidence has risen considerably. After the mixed confusions and striking successes surrounding the Gemini 6 and 7 missions, NASA has dedicated itself even more strongly to the remaining elusive objective: a successful docking with the Agena Target Vehicle. Achieving this goal has become the focal point for Gemini 8. A successful docking of two separate vehicles in orbit has yet to be attained, but the long-range objectives of our space program demand its completion. The Gemini capsule in orbit is a highly maneuverable spacecraft with numerous capabilities. But essentially it remains a sophisticated orbital device because it is equipped with small maneuvering engines only. Coupling the capsule with the Agena will place a 16,000-pound thrust engine at its disposal. Gemini will no longer be an advanced Mercury, it will then be an extremely powerful spacecraft with the ability to thrust itself and its manned cargo across the face of the earth at tremendous velocity. The docked vehicles will further be able to undertake extreme orbital changes safely. The opening days of 1966 are crowded ones at the Cape and in the contractors' factories. Bell Aerosystems has uprated the Target Vehicle under their "Project Surefire", while Convair has conducted intensive studies of their Atlas boost vehicle. A total of 20 engineering alterations have been incorporated into the target's launch vehicle since the Agena failure of October 25, 1965. Failures are no longer a deterrant, they only prepare the way with determination.

NEIL A. ARMSTRONG

COMMAND PILOT/ Born in Wapakoneta, Ohio, August 5, 1930, Armstrong earned his BS degree in 1955 from Purdue in Aeronautical Engineering. He did graduate work at USC and is an ex-Naval aviator, having served in Korea with 78 combat missions. He joined NASA's (then NACA) Lewis Research Center in 1955, and accumulated thousands of hours in numerous jet aircraft, both fighter and experimental. He flew the X-15 rocket airplane at 4000 miles per hour and over 200,000 feet high. Selected by NASA as an astronaut in 1962, he served on the backup crew of Gemini 5 as command pilot. Gemini 8 will be his first space flight. He has two children, and relaxes by soaring in a glider.

DAVID R. SCOTT

PILOT/ A Texan, Scott was born in San Antonio, June 6, 1932. He graduated fifth in his class from West Point, and chose an Air Force career, which greatly pleased his father, a retired USAF brigadier general. After a four-year tour of duty in the Netherlands, he studied at MIT, where he earned his Master of Science degree in Aeronautics and Astronautics. After completing his studies he attended the USAF Experimental Test Pilot School and the Aerospace Reaseach Pilot School, logging over 4,000 flying hours. He was chosen by NASA as an astronaut in October, 1963, being among the third group of men selected for the expanding space program. Gemini 8 was his first space flight.

ARMSTRONG
SCOTT

GEMINI 8
Pre-launch

The spaceman of the comic strip era was almost invariably equipped with three trusty devices: a ray gun, a radio headset, and some kind of strap-on device which allowed him to fly, either in space or from point to point on a planet. Astronauts of tomorrow may carry a "ray gun" in the form of a compact welding device. Radio headsets are among the world of reality today, but what about the "flying device"? With the rather low-key name of Astronaut Maneuvering Unit (AMU), such a piece of equipment is currently under test by NASA (**2**). Looking like a large backpack, the AMU houses freon jets which can be triggered by the wearer. So long as the supply lasts, the astronaut can shoot himself around during an EVA. The AMU's jets are multi-directional, which means that the astronaut will be able to convert himself into a miniature spacecraft. The AMU is tested under zero-gravity conditions by astronaut David Scott (**1**) in an Air Force KC-135 flying a parabolic curve. Behind him is a dummy hatch configuration of

1

2

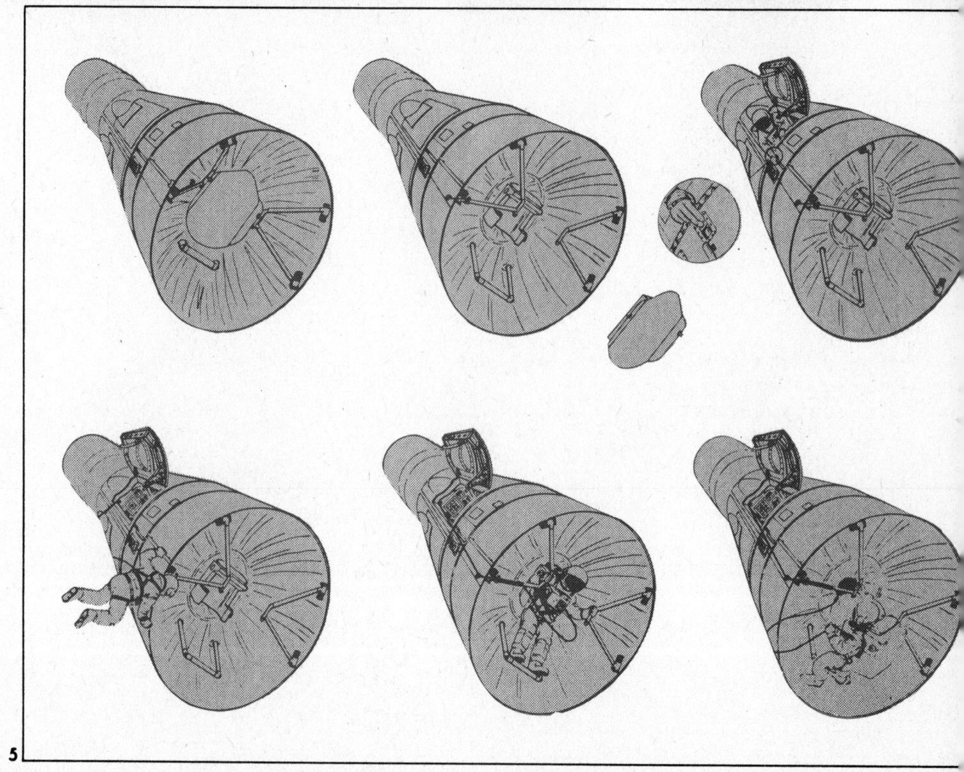
5

the Gemini capsule. Another astronaut is in the pilot's seat. In orbit, the astronaut will follow a step-by-step procedure for operating the AMU. During launch, the AMU is stowed in the aft section of the capsule's instrumentation ring (**3**) behind a jettisonable plate. Prior to EVA, a switch in the capsule is thrown, which initiates the release of the plate, simultaneously extending three "grab-irons." The astronaut then opens his hatch, and moves to the rear of the capsule. Using the handholds, he backs up against the AMU. Then, using the foot restraint, he attaches the AMU to his suit and frees it from the capsule. Easing clear of the spacecraft, the astronaut becomes a small satellite, capable of propelling himself around the capsule within the limits of his restraint tether. In early March, 1966, the GATV #5003 is prepared for attachment to the Atlas launch vehicle (**4**). All systems not protected by the hull are shrouded in heavy plastic. Assembled and shining in the morning sun, the target launch vehicle undergoes a complete checkout of all systems (**5**). A cluster of wires hang like cobwebs between the vehicle and the service tower. Atlas #5302 undergoes especially intensive tests after the failure of #5301 the previous year, which disrupted the flight plan of Gemini 6. The only new system as yet untried is a revised destruct unit. Destruction of a launch vehicle during early flight phases may become necessary by an uncontrolled flight, which would endanger lives and property. Such was the fate of that earlier Atlas.

4

GEMINI 8
Pre-launch

In the early morning hours of March 16, 1966, Mission Control is a beehive of activity (**1**). Television monitors project images of the launch vehicles into the control room as seemingly heedless Martin-Baltimore and Cape technicians watch the countdown progress on their boards. A one-day delay has been caused by overfilling the propellant tanks of the target launch vehicle. This has necessitated the replacement of the fuel tank regulator and tank relief valve. Onloading of fuel is one of the final phases of the pre-launch sequence, and surrounded by numerous safety measures. Astronauts Armstrong and Scott (**2**) share breakfast with astronauts Slayton, Anders, Cunningham and Chafee. It will be the first flight into space for both Armstrong and Scott, but their professionalism far and away counters any visible anxieties they may have as they leave for the launch complex (**3**). Armstrong's experience with the X-15 rocket plane has given him an insight into the experiences of high-speed, high-altitude flight

1

shared by few other men, whether astronauts or not. Scott earned his Master's degree from MIT with a thesis on interplanetary navigation. Two of the finest pilots in the world will soon take to the seats of Gemini 8. While awaiting ingress into the capsule (**4**), Armstrong studies one of the protective "galoshes" he will wear while entering the capsule and whenever walking; these protect the footgear on the space suit. Boots on and portable air conditioners running, Armstrong and Scott head for the capsule (**5**); a Technicolor Corp. cameraman at left films the event. Two launch complex technicians follow close behind to aid them entering the vertical capsule. Along with their primary objectives, the astronauts have a number of scientific, technological and medical experiments included in the flight plan. The single medical experiment is a post-flight examination of the astronauts' body fluids. A battery-operated, hand-held power wrench is slated to be tested during EVA. Its design will allow loosening and tightening of fasteners without transmitting any torque effect to the operator's hand. Scientific objectives include photography of zodiacal light, cloud top spectrometer observation, micrometeorite collection, nuclear emulsion and the effects of zero gravity on the growth of frog eggs. The nuclear emulsion experiment will hopefully expand our knowledge of cosmic radiation at high altitudes. This is still an important factor, about which not enough is known, when considering the longer duration of future flights.

3

5

GEMINI 8
Launch

GEMINI 8
Mission

Launch of the Agena at 9:00 AM EST signals the start of the mission; the target is inserted into an orbit of 161 nautical miles altitude and has a nearly circular path. This time Mission Control reads "go" on all systems. At 10:41 AM EST, Armstrong and Scott lift off and thunder spaceward into an elliptical 86 by 147-mile orbit. For the next six hours, the astronauts are kept busy tracking and maneuvering into a rendezvous with the target vehicle. Nine maneuvers are required, and at 5 hours, 58 minutes into the mission, rendezvous is completed. With no relative motion between the two craft, the target orbits 250 feet away from the capsule (**1**) with its engine angled 45 degrees towards the capsule. Armstrong and Scott move in closer, to 190 feet away (**2**), during the "stationkeeping" phase. As they move around the target, the docking collar passes into full view (**3**). All systems are functioning normally, and as the capsule swings around at 40 feet, the target fills the entire right-hand docking window (**4**). Tension mounts

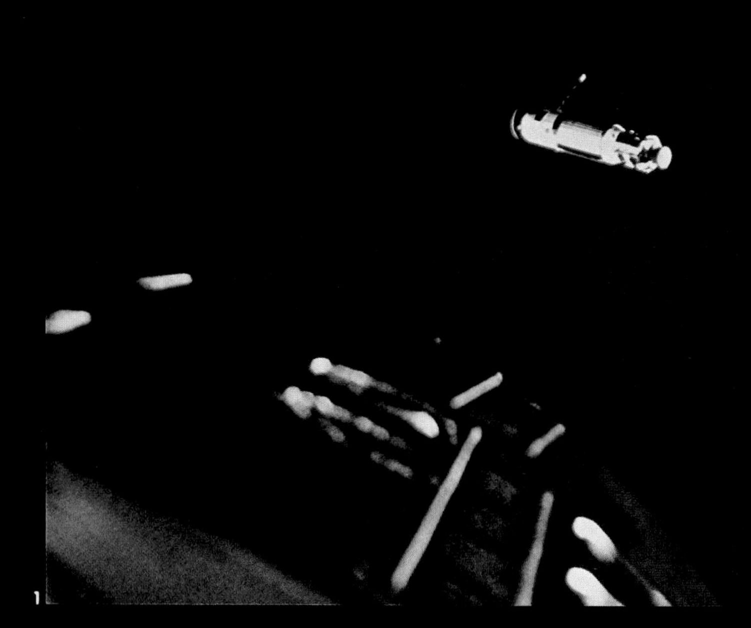

1

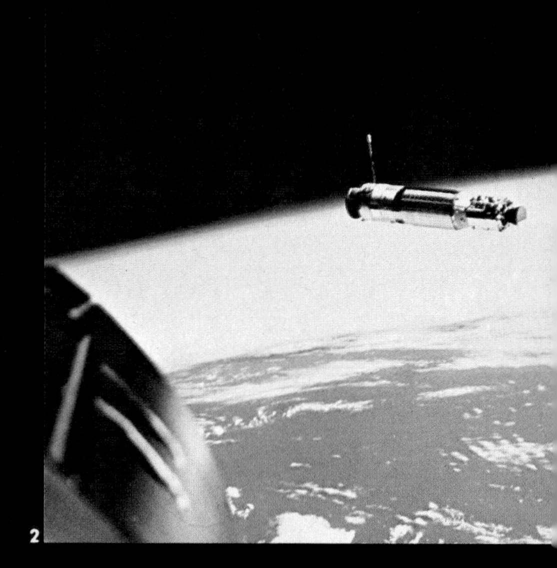

2

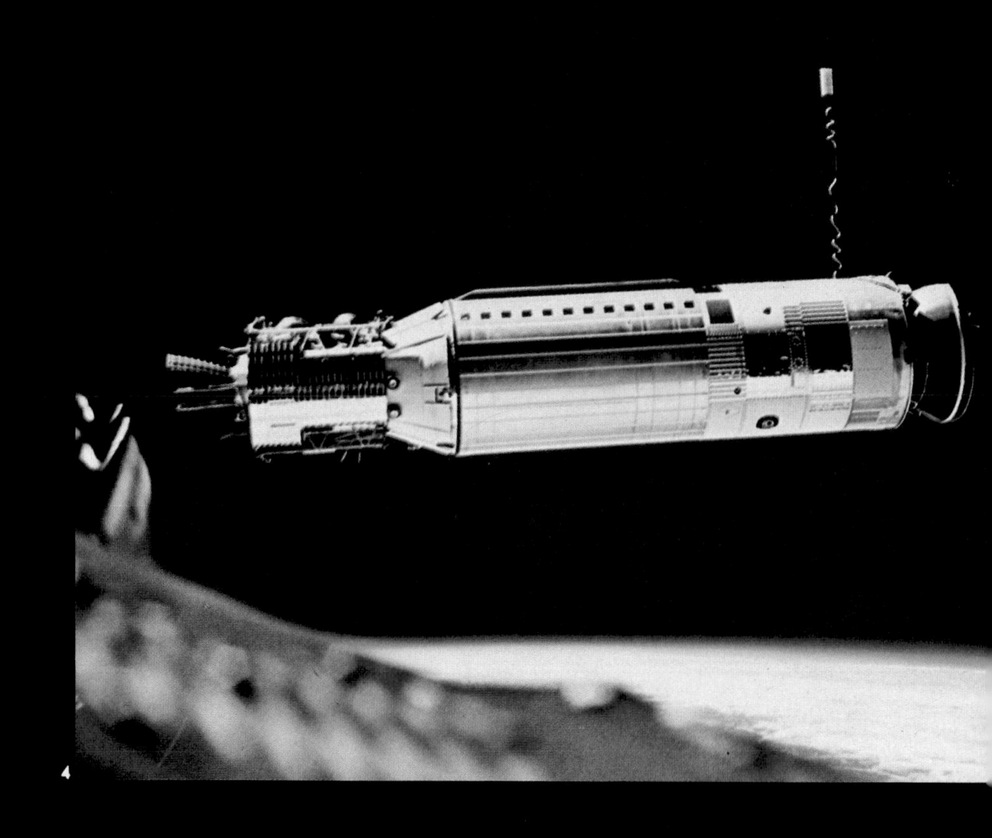

4

as the crew finally align the docking ele-ments and ease the nose of the Gemini into the Agena's docking collar (**5**). At 6 hours, 33 minutes, ground elapsed time, the histor-ic moment occurs (**6**) as the two vehicles achieve docking over the South American continent. Two separate spacecraft have fi-nally met and connected in space under di-rect manned guidance. Elated by their suc-cess, Scott and Armstrong prepare to take control of the Agena. The target responds to preliminary commands from the Gemini, and a simple yaw maneuver is completed. Sud-denly, near-disaster strikes. One of the cap-sule's roll thrusters fires and continues firing without either ground or astronaut com-mand. Its 25-pound thrust is not high, but is sufficient to begin a rolling action that in-creases in speed. "We're tumbling end over end!" exclaims the crew to Mission Control. It has been only 27 minutes since docking was achieved. Summoning every ounce of nerve and all their ability as pilots, the crew undock. Freed, the capsule contin-ues its dizzying spin, actua... tumbl... once per second. The malfunctionir... er is cut off, and the crew's only re... fire the attitude control thrusters in... tion opposite the roll. These units... marily intended to stabilize the cap... place it into correct re-entry mode; ... expend their fuel as brakes. Finally, ... ride is over. The capsule is once a... mal, but 75 percent of the re-entr... system's fuel is expended. Gemini ... in 10 hours—a flight to be thre...

3

GEMINI 8
Splashdown

Fortunately, there have been no internal disasters endangering Scott's and Armstrong's lives, but Mission Control orders Gemini 8 down, and that means now. Emergency reentry procedures are put into effect, and after a 181,000-mile flight, Gemini 8 puts on the retro-rocket brakes and splashes down in the Pacific Ocean southeast of Okinawa. Scott and Armstrong remain in the capsule for about an hour, then egress and deploy their inflatable life raft. Help is on the way in the form of the USS *Mason*, a destroyer. Having splashed down at 10:30 PM EST, the Mason arrives for the recovery at 1:30 AM EST. The capsule shows its blistered face (**1**) to the camera as it is hauled aboard. Aside from the docking achievement, little else was accomplished during the flight. Mission Control resumes handling the Agena, and after bringing it under control proceeds to fire the main engine nine times. The target responds to 5000 commands until its electrical power is exhausted after 10 days. All fuel for the attitude control system

1

2

4

is vented, and the vehicle left abandoned in a circular orbit, 220 miles high. It will be inspected by future Gemini crews. Armstrong and Scott return to Okinawa, where, despite rainy weather, an impromptu welcome awaits them from local military personnel (**3**). A band in raincoats plays for the astronauts. Above the crowd, Armstrong and Scott pose with the pararescuemen who first reached their capsule the day before. Standing between the astronauts is E.M. Neal; front, from left, L.D. Huyett and G.N. Moore. From Japan, the astronauts fly back to the more familiar surroundings of Cape Kennedy, with a stopover at Pearl Harbor's famed Hickham Field, (**4**), where another welcome awaits them, headed by astronaut Wally Schirra, who walks alongside Armstrong in coat and tie. McDonnell Aircraft concludes that the cause of the unwanted thruster operation was due to an electrical short-circuit. A similar malady on earth can ruin a TV program or stop a toaster from operating. On Gemini 8 it caused a solenoid to operate, allowing fuel to pass to the thruster. Success sprinkled with frustration has been the story of Gemini. Now NASA has two projects underway at the Cape, the second being Apollo, our grand finale to the first decade of space exploration. Hardware for the lunar voyages has been under development since the time of Project Mercury, and our industry, rising to the highest pitch of peacetime effort in history, is now delivering. Yet Gemini is not complete. We still have objectives to meet, and NASA plans.

GEMINI 9

An ill wind blows against this mission; barely three months before the projected launch date, it almost destroys itself. On February 28, 1966, Gemini 9 astronauts Elliot M. See, Jr., and Charles A. Bassett II are killed when their T-38 jet training plane crashes in rain and fog short of St. Louis Municipal Airport. It is left of center on its approach to the runway when it curves toward the McDonnell complex 1000 feet from the landing strip, striking the roof of the building housing spacecraft No. 9, the one intended for See and Bassett, as well as spacecraft No. 10, before exploding in the adjacent courtyard. Minutes later, Thomas P. Stafford and Eugene A. Cernan, backup crew for Gemini 9, land safely. All four astronauts had been en route to McDonnell for two weeks' training in the simulator. Now there are two. Stafford and Cernan, NASA headquarters announces, will fly the mission on schedule. That schedule calls for a launch date of May 17, some two and a half months from the inauspicious omen at St. Louis. It calls for rendezvous and docking with an Agena target vehicle, and for a spacewalk of perhaps some two and a half hours—the latter involving the actual assumption of a work load in space via new equipment called the Astronaut Maneuvering Unit (AMU). In recognition of these objectives the mission insignia shows an astronaut moving through the airless void of space, and an Agena rocket docked with a Gemini capsule. Originally intended for use during the Gemini 8 mission, the AMU has been moved back to Gemini 9 because of the dangers encountered on that earlier flight. The AMU will enable man to move and work in space.

THOMAS P. STAFFORD

COMMAND PILOT/ Born in Weatherford, Oklahoma, on September 17, 1930, Stafford received a BS degree from the U.S. Naval Academy in 1952. Commissioned in the Air Force upon graduation from Annapolis, he flew interceptor aircraft before attending the Experimental Flight Test School at Edwards AFB, California. There he served as chief of the Performance Branch, in addition to writing flight test manuals and basic textbooks. An Air Force Major, he has logged flying time in excess of 4500 hours. Previous NASA experience includes pilot on the back-up crew for Gemini 3, and pilot on the Gemini 6. In addition he monitored development of NASA communications and instrument systems.

EUGENE A. CERNAN

PILOT/ Born in Chicago, Illinois, March 14, 1934, Cernan earned a BS in Electrical Engineering from Purdue University in 1956, as well as a MS in Aeronautical Engineering from the U.S. Naval Postgraduate School in Monterey, California. Between degrees he was assigned to Attack Squadrons 126 and 113 at Miramar Naval Station, California. Out of more than 3000 hours flying time, almost all have been in jet aircraft. Though this will be his first space flight, in previous NASA work he has monitored the spacecraft propulsion systems and the Agena D (the Gemini target vehicle) and has served as spacecraft communicator (Capcom) in the Mission Control Center on several previous Gemini flights.

GEMINI 9
Pre-launch

The sequence of events leading to launch is as precise as the hands of man can make it, each separate portion fitting together in its own pattern of time and place. Spacecraft No. 9 and its Agena target vehicle, No. 5004, begin compatability tests at Merritt Island on March 21; the Titan launch vehicle for Gemini 9 is removed from storage and erected (**1**) at Complex 19 of Kennedy Space Center on the 24th. The spacecraft is hoisted to position (**2**) atop the launch vehicle on March 28, and power is applied on the 29th. Atlas 5303, which will launch the Agena target for Gemini 9, is erected at Complex 14 on April 4, and power is applied for the first time April 11. All is ready for the May 17 launch date. Launch and flight are normal (**3**) until about 120 seconds after lift-off, ten seconds before booster engine cut-off. At this point booster engine No. 2 shuts down, and automatic correction proves ineffective. Though stabilization is achieved, the Agena executes a 216-degree pitchdown maneuver (which leaves it point-

1

2

3

ing at the Cape) at a climbing angle of about 13 degrees above the horizon. Then ground guidance is lost. Continuing on its new trajectory with normal sequencing, GATV 5004 separates but cannot attain orbit; it falls into the Atlantic ocean some 90 miles off the Florida coast, about seven and a half minutes after launch. So the mission must be postponed, and redesignated as Gemini 9A. NASA then decides to launch the augmented target docking adaptor (ATDA) due to the loss of launch vehicle 5303 and GATV 5004. Atlas 5304, intended for Gemini 12, is removed from storage and modified to serve as launch vehicle for the ATDA. Modification is completed by May 20. Launch vehicle 5304 is erected at Complex 14 the next day, and is mated to the ATDA on May 25. On June 1, this combination is launched at 10:00 AM EST (**6**) and achieves a near circular orbit (apogee 161.5, perigee 158.5 nautical miles). The astronauts, on this first day of June, their second launch date, proceed without fanfare through the pre-launch routines. Though the failure of guidance mechanisms thus far cannot have been without its effect, they do not show it. Stafford (**4**) shoves up his sleeves in the classic act of a man getting ready to take care of business; their spacecraft will be under his command, and he will be ready. On the other hand, Cernan casually reads through the morning paper (**5**) while he is hooked and fitted into his spacesuit. He might as easily be at the barber shop as being prepared to be blasted off the earth.

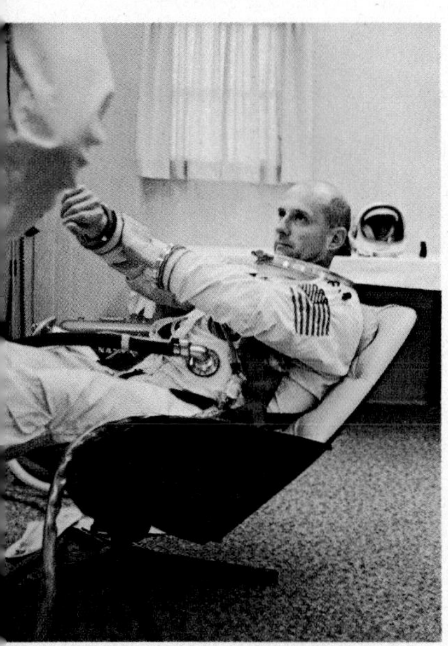

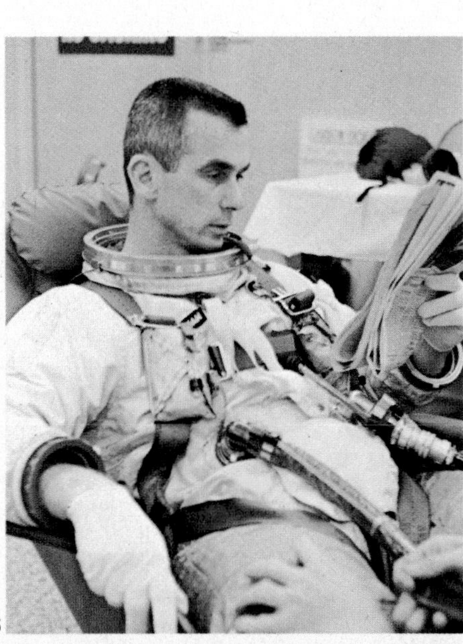

5

GEMINI 9
Launch

In the white room high atop their Titan, Cernan and Stafford prepare to enter their capsule for the second launch attempt (**1**). Determination is evident behind those visors—but a waiting McDonnell technician relays information that one hour and forty minutes after launch of the ADTA, the scheduled launch of Gemini 9A will be postponed. Before, it was the failure of their target to achieve orbit; today it is a ground equipment failure which prevents transfer of updating information to the spacecraft's computer. Stafford winces at the news, as (**2**) Cernan puts a hand to his shoulder to steady them both. All the dreams of the night before, all the hours of preparation, are vaporized in seconds. Launch is rescheduled for June 3, 1966. In early light that morning, with a grim smile at the dawn (**3**), they try for a third time, wondering if fate will provide its charm. The humor in a note from the back-up crew, Jim Lovell and Buzz Aldrin (**4**), cuts through the frustrations of this project. Finally, (**5**) at 8:39 AM the mighty Titan roars success.

1

2

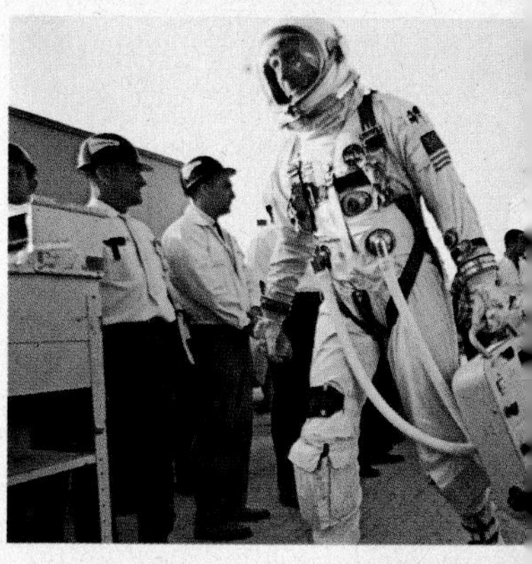

3

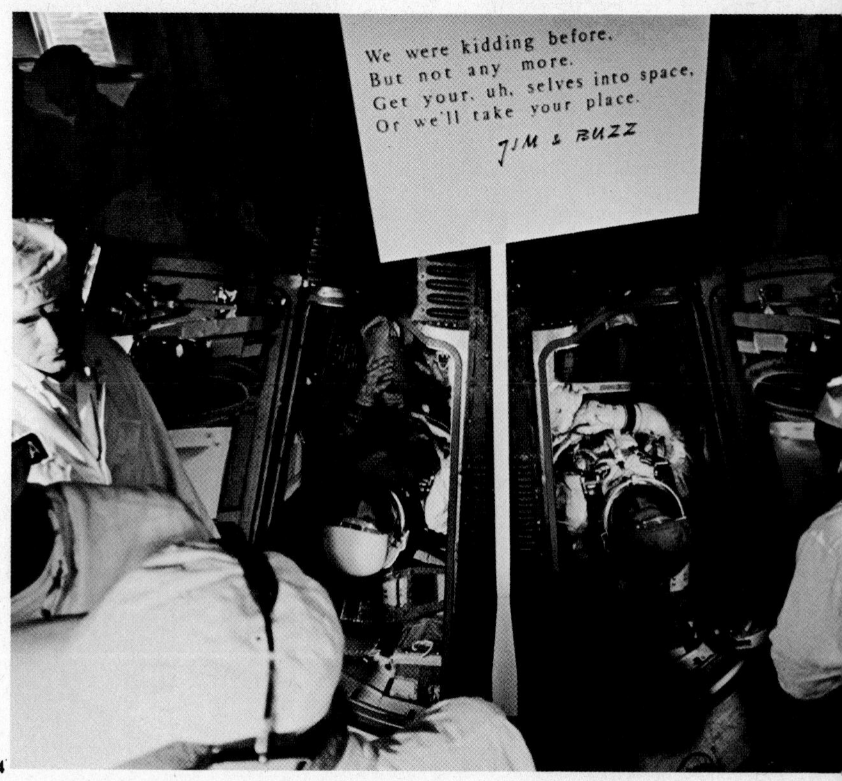

We were kidding before.
But not any more.
Get your, uh, selves into space,
Or we'll take your place.
JIM & BUZZ

4

GEMINI 9 Mission

Major objectives for this mission are rendezvous and docking with the ADTA, as well as conducting more elaborate extravehicular activities (EVA). Whirling through space to meet their intended target, astronauts Cernan and Stafford pass over that part of South America which was once the ancient Inca empire. Lake Titicaca, largest on the continent, flares blue below them, in a view encompassing Peru, Bolivia, Chile and the Andes highlands (**1**). Further on, over the mouth of the Mississippi River (**2**), they sight the river tracing its silver path from the upper left to its delta, right of center. In their third revolution, they approach the ADTA. And trouble. Again. Closing to within 70 feet of the ATDA over the coast of Venezuela (**3**), they notice the shroud which protected the nose on its flight through the atmosphere has failed to separate completely. Stafford reports to ground control, "We have a weird looking machine... It looks like an angry alligator out there rotating around." Docking with this space monster is obviously

1

2

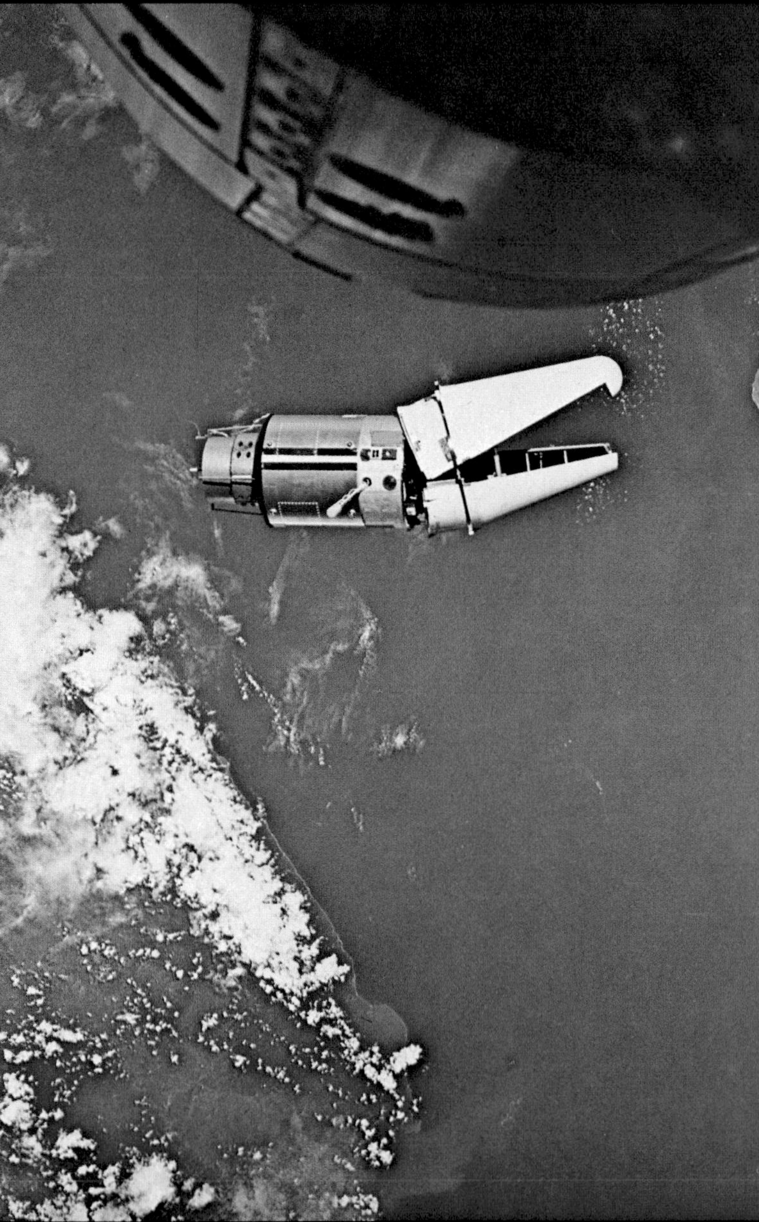
3

out of the question. However, they are able to achieve a rendezvous from above, simulating that of an Apollo Command Module with a Lunar Module in a lower orbit, and Stafford maneuvers Gemini 9A to within three feet of the ATDA at one point. Having done what they could, final separation maneuver is performed 22 hours 59 minutes after lift-off. Due to crew fatigue, the second day is given over to experiments. On the morning of June 5, the third day, Cernan prepares for his spacewalk, programmed to be a long one during which he will test the AMU with the object of moving from Gemini to the target vehicle if possible. Climbing out of the capsule at 10:22 AM EST he slowly moves to the adapter skirt at the rear of the spacecraft where the AMU is stored (**4**). But as he struggles to get into the straps, the Extravehicular Life Support System becomes overloaded with moisture, apparently because Cernan has to work harder than anticipated. His visor fogs, cutting his vision so drastically that he is unable to perform the operation. Still, he manges to capture this view of the capsule to which he is tethered, with its hatch door open to the vastness of space and its white nose bright in the sun (**5**). On advice from Stafford (**6**) he re-enters the spacecraft after two hours and five minutes of exposure to space. The rest of the third day is spent on experiments—not the least of which is remarkable space photography (**7**), such as this view of Baja California and Sonora, Mexico. As with all photos taken in airless space, the clarity is striking.

GEMINI 9
Splashdown

The AMU which has given Cernan so much trouble on this mission is a rectangular aluminum backpack weighing 166 pounds fully loaded, including 24 pounds of hydrogen peroxide as fuel. It houses twelve small thrusters mounted at the corners (it is 32x22x19 inches) and a battery-powered UHF transceiver. It and the adapter section which houses it are jettisoned before re-entry. The first high altitude drogue chute is deployed at 50,000 feet, followed by pilot parachute deployment at 10,600 feet and main chute deployment at 9000 feet. Despite the problems which have plagued this effort—the inability to dock with the ATDA or to use the AMU are major failures in themselves and furthermore prevent completion of several secondary objectives including most of the experiments—Gemini 9A racks up a few unqualified successes. It accomplishes history's fastest rendezvous of space vehicles, as well as its longest spacewalk. To these now is added an utterly flawless re-entry. AT 9:00 AM EST on June 6, 1966,

1

2

3

4

5

the Gemini 9A splashes down in the western Atlantic (27°52'N, 75°W) some 345 miles east of the Cape, after 45 revolutions of the globe and 1.2 million miles of travel representing 144 man-hours in space. The brief flutter of a parachute (**1**) marks the splashdown site, about one-half mile short of the aiming point and only three and a half miles off the bow of the USS *Wasp,* the main recovery ship. The Air Rescue Service quickly flies in to affix the floatation collar (**2**), and (**3**) astronauts Cernan (left) and Stafford pop out to chat with one of the frogmen—seemingly as casual as though crews of two pleasure craft had encountered one another on a lazy weekend cruise. More frogmen join them and they elect to remain aboard as their craft is transferred to the *Wasp.* Stafford gives the positive thumbs-up sign to the hovering Sikorsky HSS-2, while Cernan reclines (**4**) against the hatch door. Noticeably absent in this photograph is the white nose of the orbiting spacecraft, the rendezvous and recovery section, which is jettisoned after pilot chute deployment. Fifty-three minutes after landing, the spacecraft is hoisted aboard the *Wasp,* (**5**) where the astronauts are greeted by John Stonesiser, NASA engineer, with whom they share a joke—perhaps about the relative hazards of space and earthbound existences, for the engineer has had some trouble with his finger. Later (**6**) Stafford and Cernan are further congratulated by the President, making his traditional telephone call of welcome. Bassett and See would have been proud of their backup crew.

GEMINI 10

As Project Gemini enters its final flight phases, the work load for both ground personnel and astronauts has increased considerably. Technical frustrations surrounding earlier missions have made the achievement of Gemini goals even more important. The successes and failures of Gemini 9 have demonstrated our capabilities in space, but also have highlighted problem areas in our inventory of space hardware, primarily in the unmanned target vehicles. Much satisfaction can be held on our part in that we have now demonstrated a clear technical superiority over the Soviet space program, at least in the manned phase. But aside from international competition, our ultimate manned goal, the moon, still seems a long way off, although the distance is narrowing with each day. Hardware for Project Apollo is already under development, and the present functional capabilities of Gemini must be utilized immediately in order to phase into Apollo. That immense project of the near future will rely heavily on lessons learned from Gemini, paramount among which are rendezvous and docking techniques as well as tracking and communications. Astronauts seasoned by Gemini will doubtless be able to make the transition into Apollo with a minimum of extra effort. Aside from meeting these technical objectives, Gemini 10 will carry out a total of 16 secondary experiments, of which 15 are devoted to science and technology, and one is a medical experiment. The prime crew for Gemini 10 will be John Young and Michael Collins. Young is a veteran of the first manned Gemini flight, Gemini 3, while for Collins it will be his first flight into space. Collins' specialty is pressure suits.

JOHN W. YOUNG

COMMAND PILOT/ A native of California, Young was born in San Francisco on September 24, 1930. He graduated from Georgia Tech with a degree in Aeronautical Engineering; a degree earned with highest honors. He entered the Navy, and presently holds the rank of Commander. From 1959 to 1962 he served as a test pilot, and later as program manager for the F4H weapons system. In 1962, he set world time-to-climb records in the Navy's F4B Phantom jet. Young was among the second group of astronauts selected by NASA in 1962, and is already a veteran of space, having served alongside Gus Grissom as pilot on the first manned Gemini flight, Gemini 3. He is the father of two children.

MICHAEL COLLINS

PILOT/ Born in Rome, Italy on October 31, 1930, Collins graduated from the United States Military Academy with a Bachelor of Science degree. Following his graduation, he chose an Air Force career, currently holding the rank of Major in that branch. He served as a flight test officer at Edwards AFB in California, logging more than 3000 hours in jet aircraft, primarily fighters and interceptors. Collins was among the third group of astronauts selected by NASA in 1963. Aside from devoting himself to the astronaut training program, he has specialized in the study of pressure suits. He was backup pilot for the Gemini 7 mission. Collins is married and has three children, two daughters and one young son.

GEMINI 10 Preparation

Pre-launch training for Gemini crews has reached peak levels prior to Gemini 10. Considerable time is spent learning docking techniques in a simulator at Langley Research Center in Hampton, Virginia. The crews' studies and orientation now include extensive travels to several of NASA's specialized training and research centers. Contractors, too, will "borrow" an astronaut for consultation or training at a particular manufacturing site. One phase of space flight that has become very important is photography, and the astronauts have spent time familiarizing themselves with camera equipment and procedure. Here, Mike Collins (**1**) looks over an alloy case with shock-packed photo gear that will go with him into space. Astronaut/Assistant Flight Director Deke Slayton looks on. On June 7, 1966, Gemini Launch Vehicle #10 is removed from controlled access storage and erected at Launch Complex 19, while the spacecraft (**2**) is hoisted into place two days later as the Titan is undergoing Subsystems Reverification Tests. Tests on in-

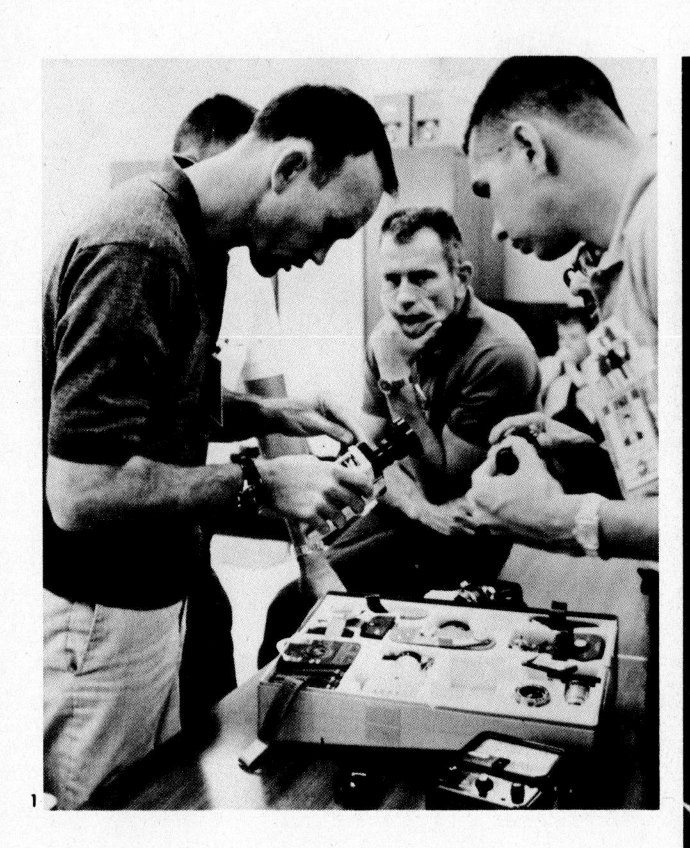

1

2

dividual components are complete by June 16, and preparations are made immediately for integrated spacecraft/launch vehicle testing. Mike Collins (**3**) has made the study of pressure suit systems his specialty. Here he is testing the fit of the Gemini EVA suit with the help of two technicians. They are connecting oxygen lines. Collins then proceeds to check out the Astronaut Mobility Unit chest/back pack in conjunction with a camera (**4**). The AMU will be attached to the ballistic face cover of the capsule, and the astronaut will have to swing himself from the side of the capsule around to the forward face in order to affix the unit to his suit. Gene Cernan experienced difficulty with this EVA sequence on Gemini 9, and was unable to complete the maneuver. Thus, the AMU remains something of an unknown. Nonetheless, the astronauts have experienced using it during short weightlessness periods in Air Force KC-135's (**5**). A fully suited astronaut free falls in an airplane during a maneuver known as a ballistic arc, sort of an "outside loop" which throws his weight against gravity from a climbing approach. He is aided by flight crew members. Atlas #5305 target launch vehicle, mated with Agena #5005 second stage, is erected on Launch Pad 14 on June 15 and systems checkout begins immediately. A fuel tank leak is corrected, and NASA works for a mid-July launch date as all pre-launch tests follow up without further problems. Pre-flight activities are completed with the Simultaneous Launch Demonstration on June 24.

3
4
5

GEMINI 10
Launch

July 18, 1966—launch day begins for Young and Collins with a good breakfast (**1**), shared with members of the launch team. Shortly thereafter, they suit up and head for the waiting spacecraft (**2**). Launch activities begin with the lift-off of the Atlas-Agena target vehicle (**3**) at 3:40 PM. The crew awaits while Mission Control tracks and verifies that the target vehicle is inserted into a 162 by 157-mile orbit. The mission profile calls for an added bonus—rendezvous with the Agena target vehicle "left over" from the Gemini 8 mission. In this multiple exposure (**4**), we see the service tower lower, the Titan's twin engines ignite and the launch vehicle lift off from Pad 19. Launch occurs smoothly at 5:20 PM, some two hours after the target vehicle has left. Spacecraft #10's objective is a very elliptic orbit of 86 by 145 nautical miles, which will initially place Young and Collins approximately 1000 miles from their target Agena. They will close the gap, rendezvous and dock, carry out joint maneuvers, then go off to the Gemini 8 Agena.

1

2

3

GEMINI 10
Mission

At separation from the second stage, speed of the Gemini capsule is 25,711 feet per second, the small thrusters firing at separation adding an additional 27 feet per second. Now Young and Collins begin a group of maneuvers to close in with the orbiting target vehicle. At four hours, 26 minutes Ground Elapsed Time, the astronauts have their first visual contact with the GATV, and at five hours, 23 minutes, the hoped-for rendezvous is achieved. The unmanned vehicle (**1**) is seen by Mike Collins from the port window, with the docking ring to the left. Carefully, Young and Collins swing the capsule around, matching their course and velocity with that of the target, and move in. John Young has this view (**3**) from the command pilot's seat. Unfortunately, only 380 pounds of fuel are still available for the thrusters; Mission Control had hoped for 680. Several experiments are scrubbed, including a badly needed series of docking experiments. With electrical coupling completed between the two vehicles, the

1

4

5

16,000-pound-thrust Bell rocket engine in the Agena is fired over Hawaii at seven hours, 38 minutes GET. "That was really something!" exclaims Young. "When that baby lights, you can really feel it!" After the burn tests are completed, Young and Collins enter an eight-hour sleep period. Beneath them passes the southern half of Taiwan (**2**). After awakening, a further series of "bending mode" tests are run by the crew, testing the reactions of the two docked vehicles; the cigar-shaped craft is literally wagged and rolled in orbit. No serious consequences result. Finally, a series of burns brings them into a synchronous orbit with the Gemini 8 Agena target. The cabin of the capsule is depressurized, and Collins practices standing up EVA. He reports that he has no trouble moving vertically, although his suit, when pressurized, filled the cockpit space. "What a beautiful view." comments the astronaut, as he runs out his tether. An annoying rise in fume level in the suit's oxygen system cuts short the EVA, and operations are once again curtailed un- the following day. After another rest perioc Young moves in close to the Gemini 8 tar get, and Collins initiates another EVA. Mov ing across the void, he removes the mi crometeoriod detector from the Agena, anc returns to his seat. For the first time, a mar has contacted another orbiting object. Be low, a storm whirls near Gibraltar (**5**). Debris in space; Collins' umbilical stowage bag (**6** is among the unneeded items disposed o by dumping, to be destroyed on re-entry

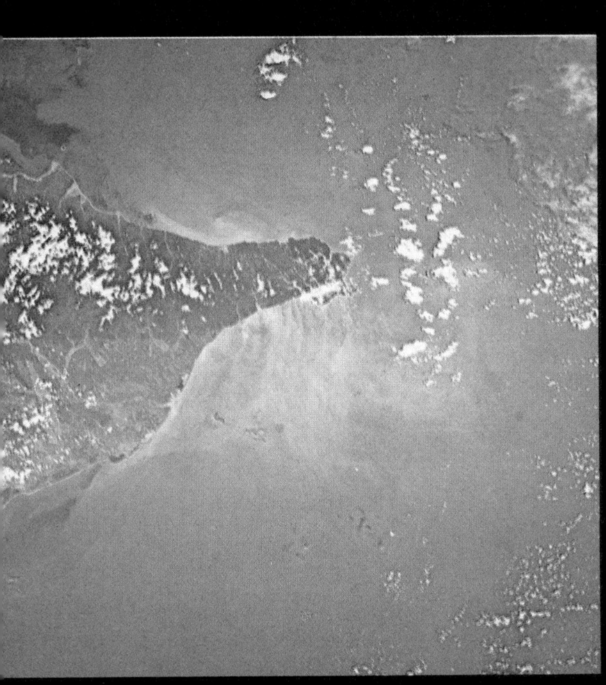

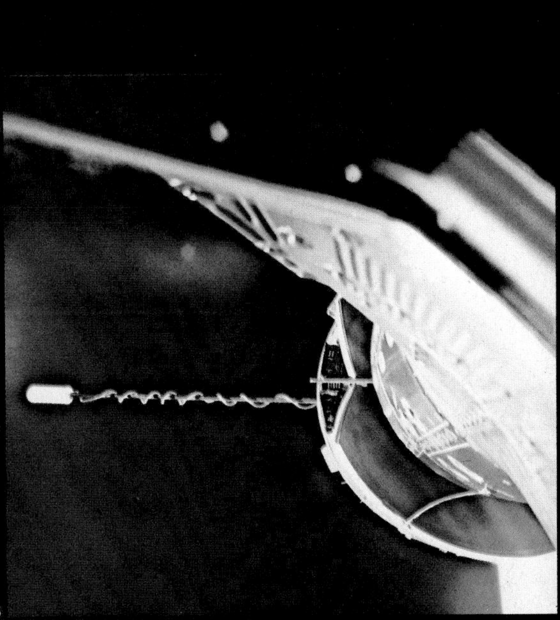

3

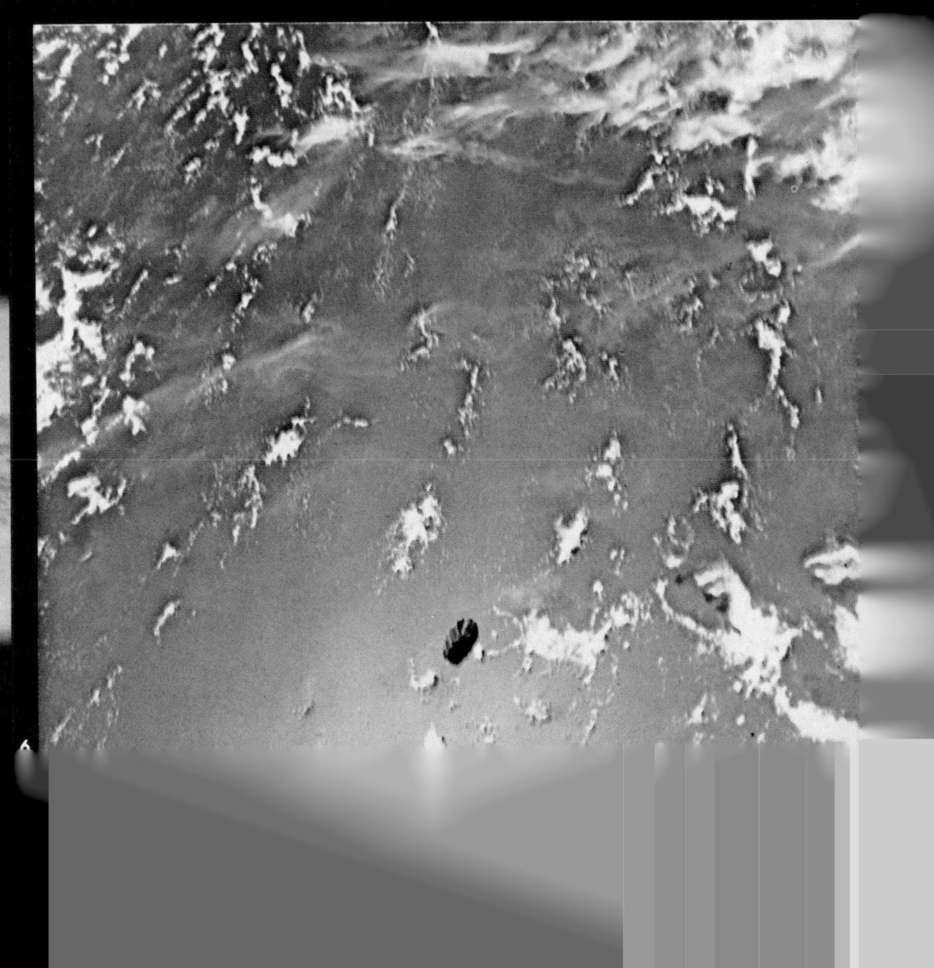

6

GEMINI 10
Splashdown

Splashdown accuracy is very good for Gemini 10, in fact, it is within four miles of the USS *Guadalcanal,* the prime recovery ship. A telephoto lens captures the spacecraft drifting seaward (**1**) towards a 4:07 PM splashdown. Twenty-two minutes later, Young and Collins step down onto the flight deck of the *Guadalcanal* (**2**) and are greeted by a small cluster of crewmen and newsmen, all seemingly going mad with cameras. And why not? The 43-orbit flight lasted 70 hours, 47 minutes and had a duration of over 1,200,000 miles. The astronauts speak to the crew and nation from the deck (**3**) under the camera's eye while a Marine color guard holds the national ensign. Still orbiting, and with life left, the target vehicle is directed into three orbital maneuvers by ground controllers. Its primary propulsion system is fired, redeveloping the orbit into a 750.5 by 208.6 nautical-mile path in order to determine the effects of temperature on the vehicle. When no significant problems are noted, the engine is fired again to place the

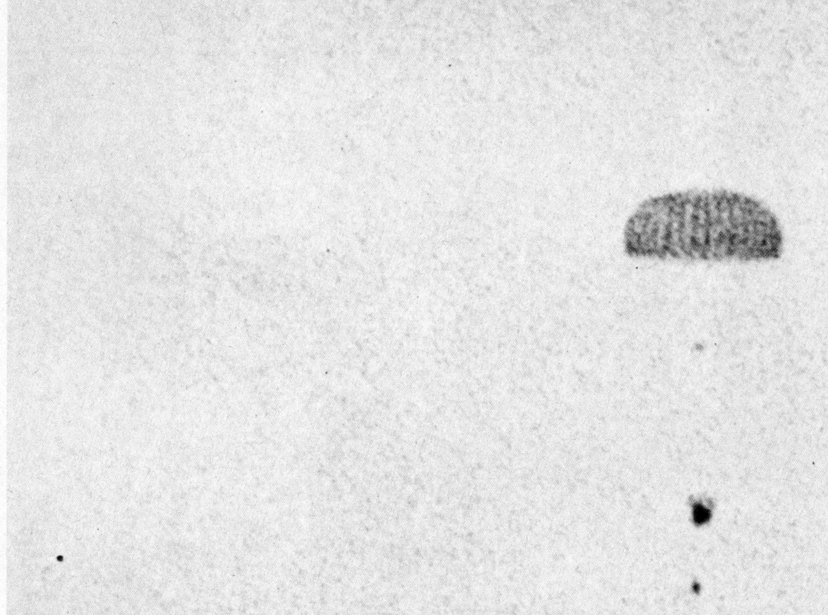

1

vehicle in a circular orbit 190 nautical miles high. In finalizing this orbital path, the secondary propulsion system is also utilized by ground control. During its active orbital life, the GATV received and carried out 1700 commands, 1350 of which were from ground control. In its present low orbit, target vehicle #5005 will be maintained for possible use on the upcoming Gemini 11 mission, as was the Gemini 8 target by Young and Collins. Gemini 10 can definitely be called a success. Collins has made history by being the first man to contact a separate vehicle in space during his EVA. The onboard experiments should prove successful, despite the high consumption of propellant which curtailed certain other operations from being carried out. Below decks on the *Guadalcanal* a reception is held for Young and Collins. A dummy capsule received this humorous cartoon (**5**) on its "ballistic face." NASA, having undergone so much trouble in previous missions; trouble with the docking procedures on Gemini, is at last caught squarely between the GT-10 capsule and the heretofore elusive Gemini Agena Target Vehicle! Good going, Young and Collins. During the cake cutting, Collins hands a nice big piece to 12-year-old Billy Doyle (**4**), who represents 41 children of shipboard Naval personnel at the celebration ceremony. The cake is a simple representation of the *Guadalcanal* sailing on the ocean. John Young, enjoying himself, ignores a microphone at his left to say a few words to the bright-eyed and awestruck youngster.

5

GEMINI 11

Now that the flight of Gemini 10 has been completed, program personnel anticipate few if any problems during the remaining two missions. With the flawless rendezvous and docking maneuvers performed by astronauts Young and Collins, the greatest obstacles have been met, or so it seems. The problem of umbilical extravehicular activity (EVA) remains to be conquered, however. Gemini 8 had been terminated before its scheduled EVA could take place, and Gene Cernan's exhausting struggle with the Astronaut Maneuvering Unit during Gemini 9 concluded his 2 hour, 9 minute EVA. Although Mike Collins had been able to remove a meteoroid collector from the Agena during Gemini 10's mission, his brief EVA had not gone as well as desired. But spirits are high at the Cape, and Gemini 11 is regarded as the mission which will provide the necessary answers about extended space walks. The three-day flight of Gemini 11 is scheduled for launch on September 10, 1966, with astronauts Charles Conrad, Jr. and Richard F. Gordon, Jr. as crew. Neil Armstrong and William Anders are the backup team. The prime objective of Gemini 11 will be a direct-ascent rendezvous, that is, to rendezvous and dock during the first orbit, leaving the crew free to conduct their 12 scheduled experiments during passive attitude stabilization. Two of the experiments—dim light and earth-moon libration photography—are new; four others are included that were previously assigned to Gemini 8, but could not be carried out because of its early flight termination. This is the most ambitious of the Gemini flights to date, and is expected to tie up many loose ends before Gemini 12.

CHARLES CONRAD, JR.

COMMAND PILOT/ Navy Captain "Pete" Conrad was selected as an astronaut by NASA in September 1962. A native of Philadelphia, the 36-year old Conrad entered the U.S. Navy following graduation from Princeton University in 1953 with a Bachelor of Science degree in Aeronautical Engineering. After completing Navy Test Pilot School, he served as a test pilot, flight instructor and performance engineer at Patuxent River, Maryland, logging over 6000 hours in the air. Pete has spent over 190 hours in space as pilot of the Gemini 5 flight, experience he will put to good use during the Gemini 11 mission. He and Jane Conrad have four children, and Pete is another astronaut who is an auto racing fan.

RICHARD F. GORDON, JR.

PILOT/ Also a USN Captain, Dick Gordon was born October 5, 1929 in Seattle, Washington. Receiving a Bachelor of Science degree in Chemistry from the University of Washington in 1951, and his naval aviator wings in 1953, he was subsequently assigned to a fighter squadron stationed at Jacksonville, Florida. After attending Navy Test Pilot School at Patuxent River, Maryland, he served as a test pilot, flight instructor and assistant operations officer, accumulating more than 4200 hours flight time. NASA selected him for astronaut training in October 1963. The father of six children and an avid enthusiast of water skiing and sailing, Dick was backup pilot for Gemini 8, but has not yet been in space.

CONRAD GORDON

GEMINI 11
Pre-launch

Only four days after Gemini 10 roars off the pad at Cape Kennedy, Gemini launch vehicle (GLV) 11 is removed from storage and erected at Complex 19. Sub-systems testing begins at once. On July 28, the Atlas target launch vehicle (TLV) 5306 is erected at Complex 14. That same day, the Gemini space capsule is hoisted atop the Titan II; three and one-half weeks later, the Agena is connected to TLV. By the end of August, all testing has been completed and Gemini 11 is now ready for its September 9 launch.

Meanwhile, Conrad and Gordon wind up their pre-flight training and as command pilot for the mission (**1**), Pete discusses the flight plan with Alan Shepard, Chief of the Astronaut Office, while security officer Charles Buckley, Jr. looks on. As a final space suit check (**2**) is carried out, Dick Gordon considers the confinement he is about to undergo for 72 hours. Mission simulation exercises (**3**) now occupy their time and involve the backup team of William Anders (extreme left) and Neil Armstrong

1

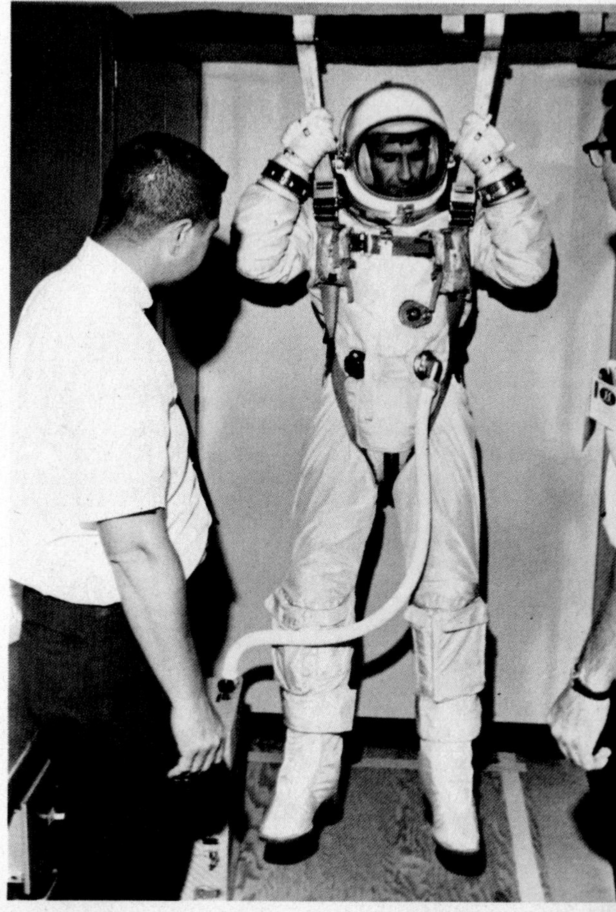

2

3

(extreme right) as well as the primary crew, Gordon (second left) and Conrad. The waiting GLV 11 stands ready for launch, its mighty engines (**4**) capable of generating over five million horsepower. Translated into rocket engine terminology, the two stages of the Titan II produce 530,000 pounds of thrust. When you recall that the Redstones which put Alan Shepard and Gus Grissom into suborbital flights generated only 78,000 pounds, and that the Atlas missiles used during Project Mercury delivered a total 360,000 pounds of thrust, Conrad and Gordon will leave the earth with a considerable kick in the pants. As command pilot, Pete Conrad's space suit (**5**) is composed of five layers and weighs only 23 pounds; Dick Gordon's suit (**6**) differs in that it has special thermal protective cover layers for EVA activities. Both suits will automatically pressurize to 3.7 pounds per square inch (psi) to provide pressure and breathing oxygen when the Gemini cabin is depressurized. Gordon's helmet is equipped with a special inner/outer visor for EVA work. The inner visor will provide impact and micrometeoroid protection, while the outer one has been gold-coated to protect his eyes from solar glare. As the two await blast-off on September 9, all systems are "Go" until discovery of a pinhole leak in the Stage I oxidizer tank of the Titan II causes a postponement of the launch to the following day. Down comes the elevator (**7**) and Conrad and Gordon walk dejectedly down the ramp to the van that will take them back to the suiting trailer.

5

7

GEMINI 11
Launch

The second attempt to launch Gemini 11 ends abruptly when an apparent malfunction in the Atlas-Agena autopilot is discovered. The flight is rescheduled for September 10. And so the two astronauts arise early on the appointed day (**1**). Sitting with Alan Shepard (right), Chief of the Astronaut Office, Conrad (left) and Gordon (center) discuss the mission over breakfast. While the two men are suiting up, the Atlas-Agena target vehicle (**2**) lifts off from Cape Kennedy's Complex 14 at 8:05 AM EST, some 97 minutes before the astronauts are scheduled to follow. Once suiting up has been completed (**3**), Pete Conrad leads the way down the steps from the Complex 16 ready room, followed by Dick Gordon and Alan Shepard. The two eagerly climb out of the van and walk up the ramp (**4**), for that third trip to their waiting spacecraft. Everything seems to be going according to plan this time. The countdown ends, the Titan's engines roar to life, and Gemini-Titan 11 suddenly lifts off (**5**), carrying the two men on their way at last.

1

2

3

4

UNITED STATES

GEMINI 11 Mission

Launched at 9:42 AM EST, Gemini 11 achieves its direct-ascent rendezvous 94 minutes after lift-off, using the onboard computer and radar equipment. An onboard camera (**1**) captures Pete Conrad hard at work as he completes the five maneuvers required to rendezvous with Agena. Once docking is completed, Conrad makes use of the Agena's restartable propulsion system to extend their orbit to a record-breaking 850-mile high point. After successfully docking/undocking four more times, the cabin is depressurized and Dick Gordon opens his hatch. Standing up, he begins the various photographic experiments. When this stand-up EVA is completed, the hatch is closed and the cabin repressurized; then Conrad returns the Gemini-Agena to a near-circular orbit. The next task is to connect a 100-foot dacron tether between the Gemini docking bar and the docked Agena during an umbilical EVA. At 24:02 Ground Elapsed Time (GET) into the mission, Gordon transfers his spacesuit cooling from the capsule's oxygen

1

2

4

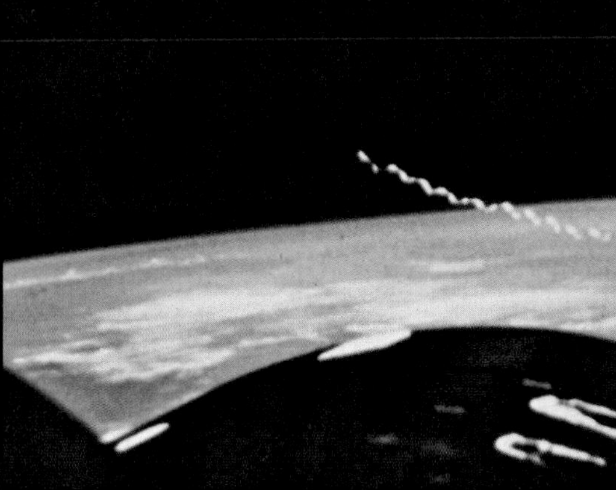

5

system to his Extravehicular Life Support System (ELSS) chest pack and just before opening his hatch a second time, reaches up to pull his EVA visor into place. Finding that the visor is stuck, Dick struggles to pull it into position and as his ELSS cannot handle the cooling load—it is designed to work best *outside* the spacecraft—Gordon soon finds himself drenched with perspiration before he manages to position the visor and open his hatch. Tethering the Agena is expected to be no more than a five-minute task, with Dick straddling the small end of the Gemini capsule as he would a horse. But he keeps floating up and away from the docking vehicle, and 20 minutes pass before the connection is finally secured. During this time, Gordon continues to perspire and his ELSS cooling system overloads. Moisture forms inside the helmet's visor, making it difficult if not impossible to see at times. Exhausted by the strenuous tethering routing (**2**), he returns to the cabin and the EVA is terminated by Mission Control. This is the second EVA cut prematurely short due to unanticipated astronaut exertion. Once tethered (**3**), Agena is deployed and the tether pulled taut to show how several spacecraft can be kept from drifting apart in space. At 53 hours GET (**4**), the tether is released and Gemini maneuvers into position (**5**) for a final rendezvous. Passing over India and Ceylon (**6**) and looking north at 540 nautical miles above the earth, the Arabian Sea (left) and the Bay of Bengal (right) make a striking picture as Gemini 11 prepares re-entry.

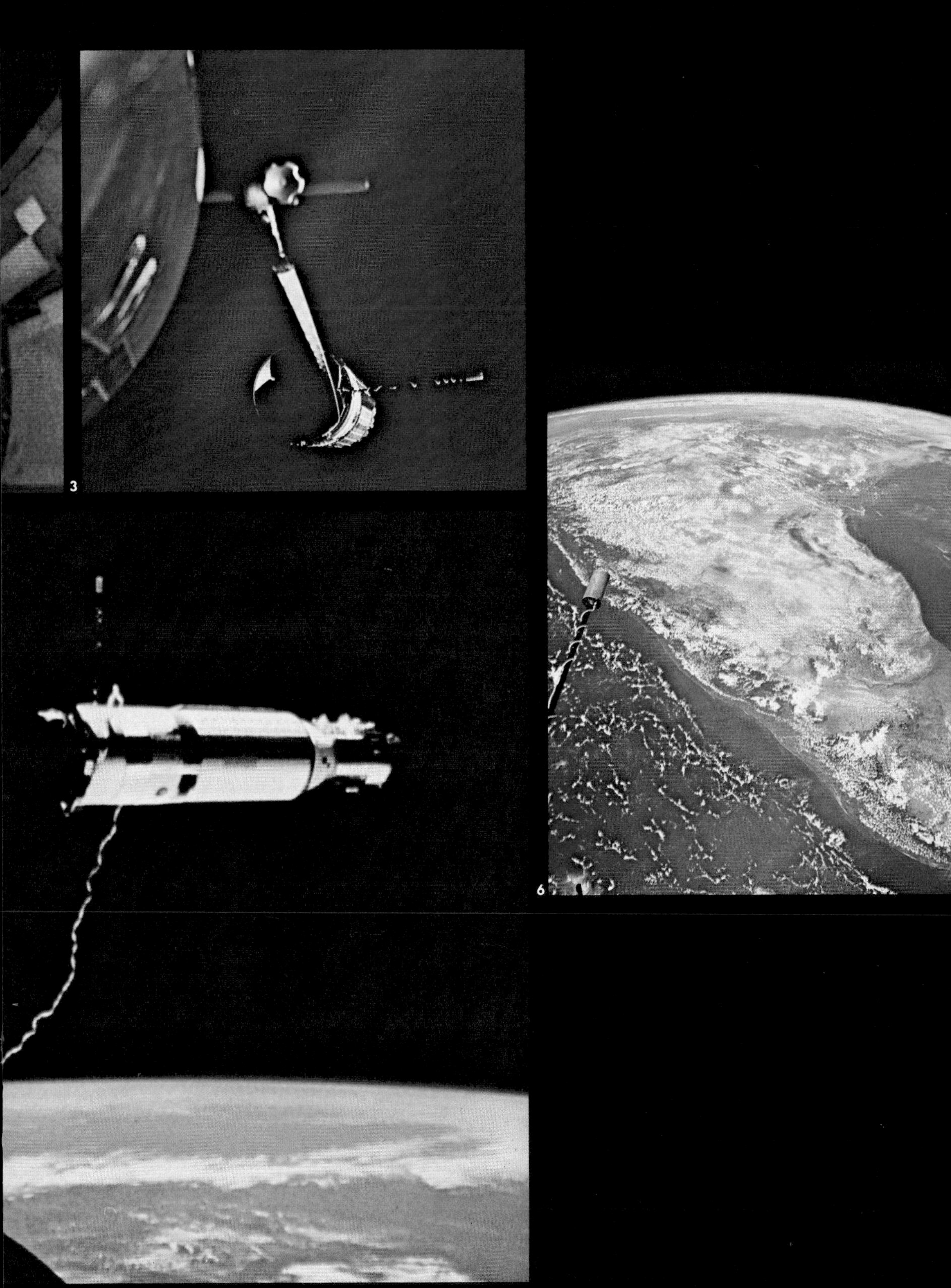
3

6

GEMINI 11
Splashdown

Once the final rendezvous is accomplished at 66:40 GET, Conrad undocks from the Agena and positions the Gemini spacecraft for re-entry. When the time comes to begin the re-entry procedure, he taps a button and an explosive device removes the adapter section's lower half, exposing the four retro-rockets which are fired in a ripple pattern to brake the capsule out of orbit. Guided by its computer-controlled autopilot (an automatic device used for the first time on this flight), the Gemini 11 spacecraft begins to slow down and fall toward the earth. Entering its long and rather flat re-entry plane, the rest of the adapter section is now jettisoned. The astronauts make their return to earth in the same position that they left it—flat on their backs. As the re-entry module reaches an altitude of approximately 10,000 feet, a small stabilization parachute opens and at 8000 feet, the 85-foot main parachute is released (**1**), lowering the capsule toward the ocean at a speed of about 20 miles per hour. During their descent, Conrad

1

2

3

and Gordon have time to reflect upon their most successful of all Gemini missions to date, and the setting of 11 new records in space. In their 71 hour, 17 minute flight, the two men circled the earth 44 times for a distance of 1,232,530 miles. Some 700 miles east of Miami, Florida, the prime recovery ship awaits their return. This first Gemini splashdown to be completely plotted and carried out by computer also turns out to be extremely accurate, landing the reentry module less than two miles from the USS *Guam* as television cameras capture the entire drama from the carrier's deck at 8:59 AM EST on September 15. Met by the USN frogman team (**2**), the astronauts barely have time to leave the capsule and enter the waiting life raft before the recovery helicopter arrives to pick up Gordon (left) and Conrad (center). Within ten minutes after splashdown (**3**), Dick Gordon and Pete Conrad step from the helicopter to the carrier deck and wave greetings to the assembled crew and reporters. About one hour after splashdown, the Gemini 11 re-entry module is fished out of the sea (**4**) and brought aboard the prime recovery ship for return to the Cape. Meanwhile, Conrad and Gordon are enjoying the traditional welcome-back ceremonies and making light-hearted jokes about their three record-breaking days in space. Scientists are especially delighted with the wealth of detail contained in the photographs taken by Gordon, and will study the pictures to gain new insights about our geography, geology and cartology.

4

GEMINI 12

The four-day flight of Gemini 12 will "close the books" on the Gemini program, paving the way for Project Apollo to place man on the moon. This last mission is particularly important, as it is the final opportunity to work out problems remaining in the Gemini series, especially those concerning the difficulties encountered by previous astronauts in working outside their orbiting space capsule. Solving the problems which had arisen while conducting such extravehicular activity (EVA) during earlier Gemini flights is considered so critical by NASA that it has revamped and extended the entire EVA program. Assignment of this objective to Gemini 12 will mean scratching a second attempt at using the Astronaut Maneuvering Unit with which Gene Cernan had difficulties during Gemini 9. The other primary objective is to put the finishing touches to already-developed rendezvous and docking maneuvers. In addition to these prime tasks, a series of secondary objectives has been assigned, most of which are associated with the EVA exercise. There are also 14 scientific experiments (two of which are new to Gemini) to be carried out, and six of which involve space photography. The prime crew for Gemini 12 will be James Lovell, Jr. and Edwin Aldrin, Jr. While Lovell is a veteran of Gemini 7, this will be Aldrin's first trip into space. Each man will wear a distinctive circular Gemini 12 mission emblem—a pink Gemini space capsule set against a blue background facing a clear quarter-moon and bearing their names and the Roman numeral XII. Launch date is set for Veterans' Day, November 9, 1966, from Cape Kennedy's famous "Missile Row".

JAMES A. LOVELL, JR.

COMMAND PILOT/ A Midwesterner by birth, 38-year old Jim Lovell was born on March 25, 1928 in Cleveland, Ohio. Following graduation from the United States Naval Academy with a Bachelor of Science degree in 1952, Lovell completed flight training. Numerous naval aviator assignments followed, including duty as a test pilot and flight instructor, until his selection for astronaut training in September 1962. He served as backup pilot for Gemini 4, and backup command pilot for Gemini 9. As pilot on the history-making Gemini 7 rendezvous mission of almost a year ago he presently shares the record for most time spent in space with Frank Borman. Captain Lovell and his wife Marilyn have four children.

EDWIN E. ALDRIN, JR.

PILOT/ Born in Montclair, New Jersey on January 20, 1930, Edwin (Buzz) Aldrin, Jr. entered the United States Military Academy at West Point and was graduated in 1951 with a Bachelor of Science degree. After flight training at Bryan, Texas, Aldrin flew 66 combat missions over Korea, then returned stateside for a tour of duty before joining the 36th Tactical Fighter Wing in Germany. Later attending MIT, his doctoral thesis (1963) dealt with the guidance requirements necessary for manned orbital rendezvous. Selected by NASA in October 1963, Aldrin was backup pilot for Gemini 9 and presently hold the rank of Colonel, USAF. He and Joan Aldrin are the parents of three children.

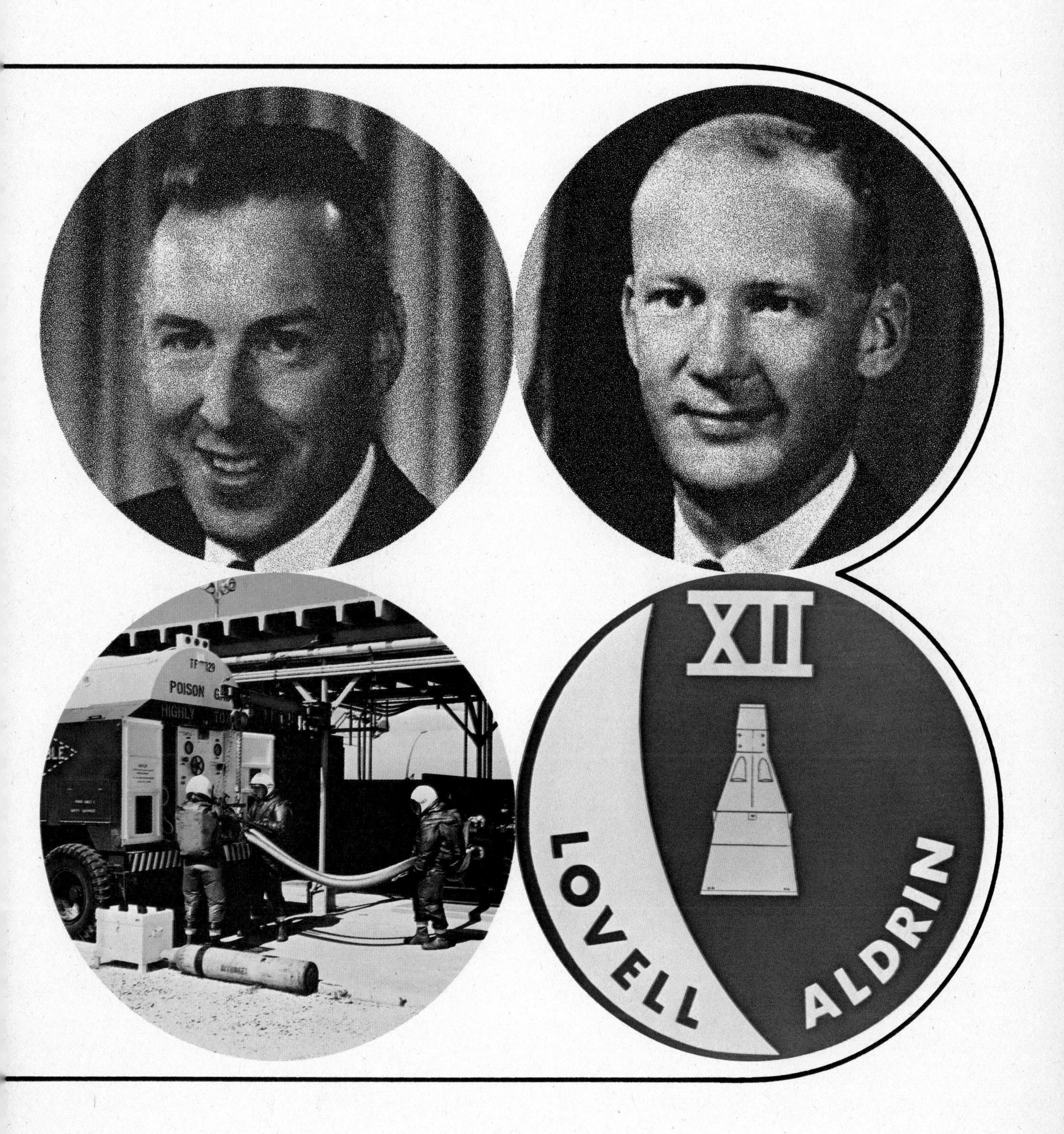
POISON
XII
LOVELL
ALDRIN

GEMINI 12
Pre-launch

This final flight will bring the Gemini program to a close, moving NASA to Project Apollo and man on the moon. Pre-launch training is rigorous and varied, and includes performing routine tasks underwater (**1**), as this activity resembles the zero-gravity condition found in space. Buzz Aldrin spends day after day underwater, slowly but steadily alternating between work and rest positions. During his rest position, Aldrin's feet are secured by "golden slippers," two step-in half-shoes that work like ski bindings to hold his feet. As the November 9 launch date approaches (**2**), world attention centers on the coming flight of Gemini 12 and astronauts Jim Lovell (left) and Buzz Aldrin. U.N. Undersecretary for Special Political Affairs Ralph Bunche (right) and the U.N. Committee on Peaceful Uses of Outer Space tour Cape Kennedy, extending greetings to the two men on behalf of the world peace body. Meanwhile, the Gemini/Titan launch vehicle GLV-12 is erected (**3**) at Complex 19 on September 19, 1966. After GLV inspection

1

2

3

4

on the 21st, Gemini spacecraft No. 12 is hoisted into position two days later. Subsystems Reverification Tests (SSRT) begin the same day, followed by premate verification which is completed October 3. Various other tests and validation checks proceed on schedule, and with the Simulated Flight Test on November 2, pre-launch testing is completed. At the same time, preparation of the Atlas-Agena target vehicle continues, despite complications. Atlas 5304 was originally scheduled as target launch vehicle for this mission, but was used for Gemini 9A after Atlas 5303 failed during target launch for Gemini 9. Scrounging around,' the Gemini Program Office locates an unused Lunar Orbiter launch vehicle, Atlas 5803. Analysis shows it to be acceptable and once modification has been completed, Atlas 5803 is redesignated as TLV 5307. After being refurbished and updated at Lockheed's Sunnyvale plant, Gemini Agena target vehicle (GATV) 5001 arrives at Cape Kennedy. This is the original Agena, built over 18 months ago. It was used during the first Simultaneous Launch Demonstration in July 1965, which showed that countdowns could proceed together on an Atlas-Agena target and a Gemini/Titan. Tests are completed on October 22, and GATV 5001 is moved to Complex 14, where it is mated to TLV 5307. Systems testing (**4**) ends with the Simultaneous Launch Demonstration on November 1. Countdown is to begin on November 8, with lift-off scheduled for 3:52 PM EST on November 9. Gemini gets ready for its finale.

GEMINI 12

Launch

3

Placing the Gemini spacecraft into a rendezvous position with the Agena is tricky; if the two are not close enough, or are in widely differing orbits, too much time and fuel will be wasted to close the gap. This launch is especially critical, as it must be done within 33 minutes from lift-off. Buzz Aldrin is practicing his extravehicular activity (EVA) in the simulator when a malfunction in the launch vehicle's secondary autopilot is discovered. Launch date is moved ahead one day to November 10 to allow time for repairs. A second malfunction occurs and the launch is set ahead once more to November 11. Jim and Buzz spend these two extra days working in the simulator. All systems finally check out and on launch day, the two men sleep until 10:00 AM. Their last meal is a hearty one—eggs, filet mignon, toast, juice and coffee. It takes them two hours to suit up. The countdown has begun and at T-120 minutes (**1**), Jim Lovell leads the way to the elevator that will carry him and Buzz to their spacecraft atop the Titan 2. The two men add a wry touch of humor to this final Gemini flight by wearing hand-made signs attached to their backs; Jim carries the word "the" and Buzz, "end." At T-110 minutes, their portable cooling units are disconnected. Meanwhile (**2**), the Atlas-Agena target vehicle blasts off from Complex 14 at 2:08 PM EST. Jim Lovell's boot covers are removed (**3**) and he steps into the spacecraft, followed by Aldrin. At T-250 seconds, automatic controls take over the countdown and at 3:46 PM EST, just 98 minutes after the lift-off of their GATV, Jim and Buzz follow aboard the Titan. The forces of acceleration push the two men back into their seats with increasing momentum. The first stage rocket shuts down and moments later, the second stage ignites, slamming the spacecraft toward orbit. After a short burst from the thrusters, the spacecraft separates on schedule from the second stage rocket and Gemini 12 is in orbit, some 575 miles behind the Agena. Jim and Buzz immediately remove their helmets and gloves, and set to work to check and double-check their navigational and guidance systems. Gemini now trails Agena by 360 miles. Another 55 minutes pass and further change corrections are made, as Gemini closes the gap to 161 miles. Still another correction of course follows 35 minutes later and the spacecraft's orbit is now 10 miles below and 75 miles behind the Agena. Preparing for radar lock-on, Buzz discovers that the radar lock is broken. In an alternate plan, he feeds data to the on-board computer for rendezvous

GEMINI 12
Mission

Rendezvous takes place about eight minutes later (**1**), upon completion of the braking maneuver. Docking occurs 28 minutes after rendezvous, and the two astronauts then perform a practice series of undockings. The first meal period follows, after which the Agena primary propulsion system (PPS) maneuver is to take place. This will change the high point of Gemini's orbit over the northern hemisphere from 185 to 460 miles and as a result, photographs taken of the earth's land forms will be different from those taken by Gemini 11. But the PPS does not work correctly, cancelling this mission objective. At 19:29 Ground Elapsed Time into the flight, Buzz begins the first of his three standup EVA periods, the easiest part of his schedule while in orbit. During its 2 hour, 29 minute duration (**2**), this EVA calls for him to open the hatch, stand up and complete several photographic experiments using a 70mm Maurer camera. At 39:30 Ground Elapsed Time, Aldrin and Lovell report little or no thrust available from the two orbit atti-

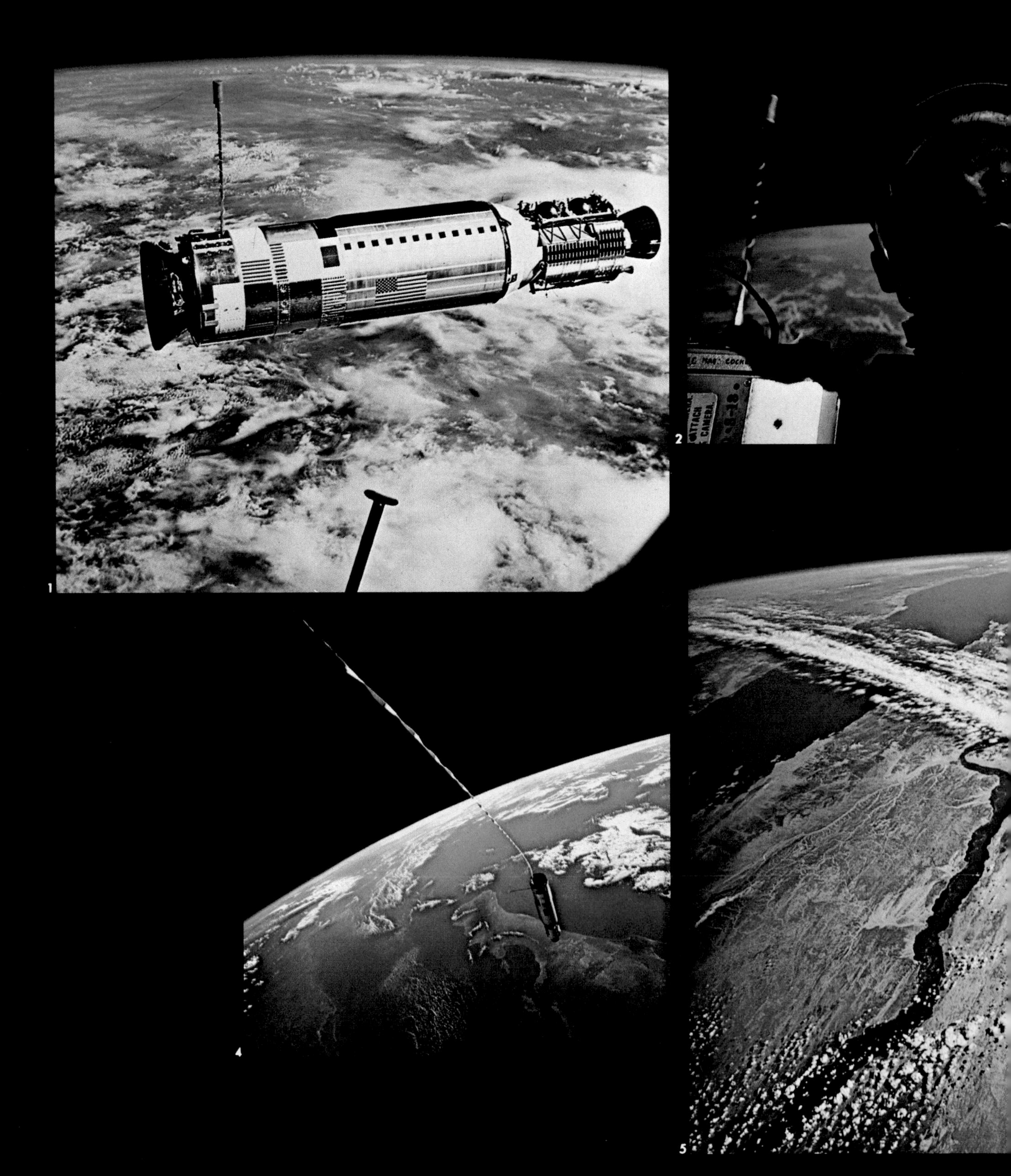

1

2

4

5

tude and maneuver thrusters. Two fuel cell stacks also fail and are shut down; two others experience a significant loss of power. During the second day in orbit, Buzz gets his chance to prove that all those hours spent underwater were worthwhile. At 42:48 Ground Elapsed Time (**3**), he leaves the spacecraft for a 2 hour, 8 minute EVA period. During this time, Aldrin attaches a 100-foot tether from the GATV to the spacecraft docking bar, installs a handrail on the Agena, works with several different restraining systems and performs various other basic tasks—clipping, bolting, racheting, etc. As he works, Aldrin occasionally looks down at the earth below—a quiet combination of brown, blue and green, with gray and white clouds floating over it. Casting a pennant into space, Buzz utters a few brief words in memory of Veterans Day and those who served their country, a tribute broadcast by Houston directly around the world. Evaluation of the tether operation begins with the crew undocking from the GATV. During the 30th revolution of the Gemini 12 mission, the Agena on tether (**4**) is photographed by Aldrin as it passes over Baja California (looking south). Although both men feel that it does its job, the tether tends to remain slack until released when the docking bar is dropped into space at 51:51 GET. On the third day in orbit, Buzz opens the hatch again and takes pictures for another 51 minutes. The jet stream (**5**) and storm clouds (**6**) passing over Mexico are some of the many weather systems he captures on film.

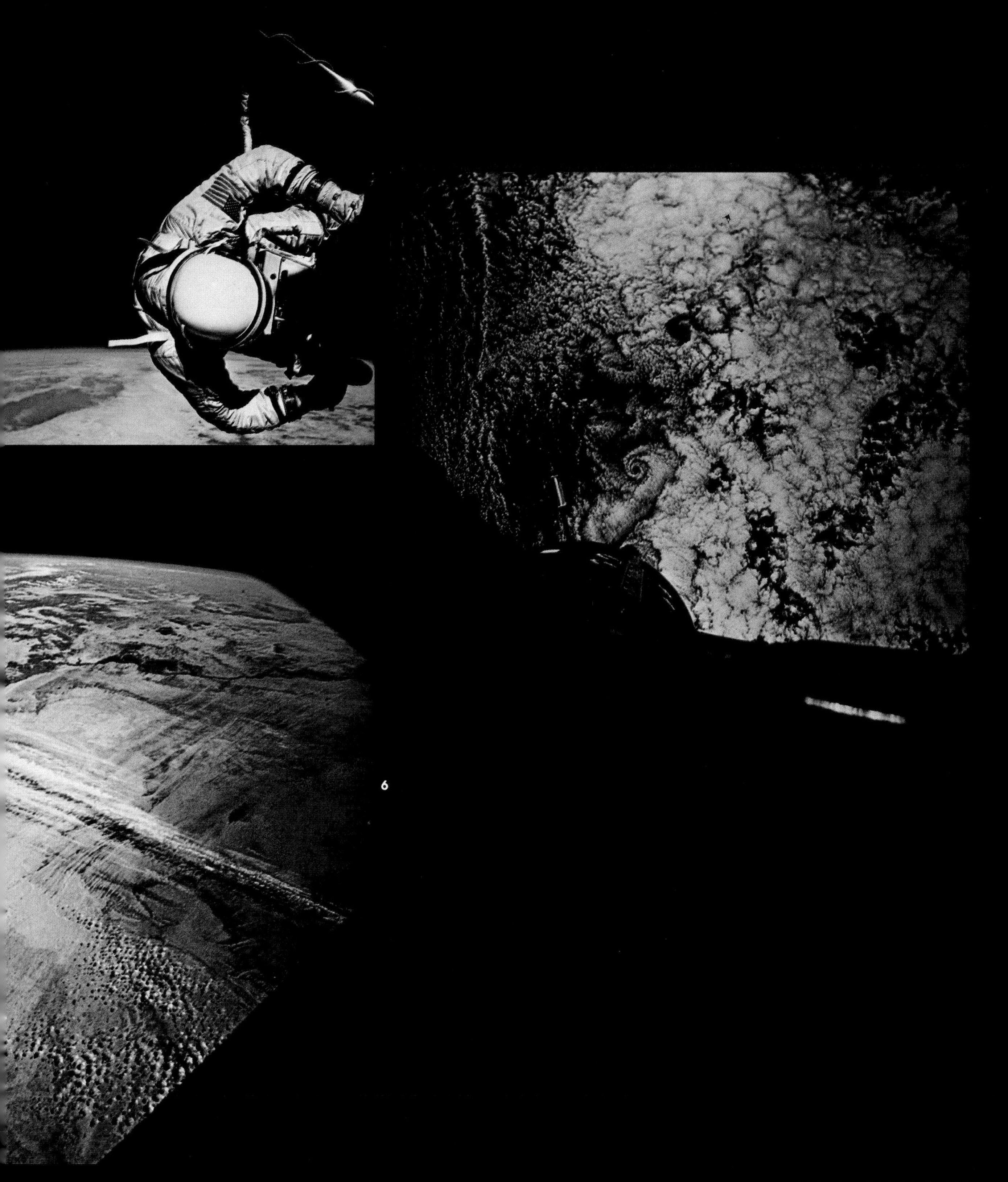

6

GEMINI 12
Splashdown

Except for the PPS maneuver and GATV parking, all objectives of the mission are accomplished. Buzz sets a record for EVA duration of 5 hours, 26 minutes, while Jim Lovell extends his time spent in space to a new record—425 hours, 10 minutes and 2 seconds. Exactly 94 hours after lift-off, Lovell and Aldrin begin the closing phase of the Gemini 12 flight which will bring them back to earth—retrofire over the Pacific near Hawaii. While this is the second Gemini flight in which re-entry and landing is automatically controlled by the onboard computer, the two astronauts can override the automatic system if necessary, but everything functions fine. Splashdown is set for a target area in the Atlantic approximately 600 miles southeast of Cape Kennedy. As its helicopters track the spacecraft's descent (**1**), the prime recovery ship USS *Wasp* stands by to welcome the returning astronauts. At 2:21 PM EST on November 15, the Gemini 12 astronauts splash down safely less than three miles from their planned landing point. Im-

1

2

3

pact is rougher than with previous Gemini landings and water begins to seep into the capsule, but within minutes (**2**), Jim Lovell is hoisted aboard the recovery helicopter, to be followed by Buzz Aldrin. This is almost the end of a long journey for both men—94½ hours in orbit, as they clocked over 1½ million miles. Only three more miles and 28 minutes separate the two from the deck of the aircraft carrier *Wasp*. Unshaven but jaunty (**3**) in their stride, Lovell and Aldrin are escorted across the carrier deck where they are to receive the official "red carpet" welcome back to earth. Enroute to Cape Kennedy (**4**), the two astronauts receive telephoned congratulations from President Lyndon Johnson and an invitation to stop by the LBJ ranch, then catch up on what has happened during their four-day absence from earth. At on-shore welcoming ceremonies for the two men later (**5**), each shares his thoughts about the success of this final Gemini mission, which paves the way for Project Apollo to begin. Buzz Aldrin injects a moment of of levity when he holds up a small banner reading "Go Army Beat Navy." Aldrin had taken it into space as a gift for Jim Lovell, a Navy officer and dedicated football fan. Chatting with friends, family, well-wishers and NASA officials (**6**), Buzz and Jim prepare for their debriefing session, then the stopover at the LBJ ranch in Texas before continuing on to New York and the hero's welcome that awaits them. Both men feel a deep sense of relief and anticipation, as man now heads for the moon!

4

6

"Where are we going? How are we going? What are we gonna do?"—these lines sum up the questions that might be asked about the Ranger program. The answers? We're going to the moon, by spacecraft, and we're going to give that old moon a good, close look, so that before man steps out on the lunar surface, he'll know whether he's going to sink to his knees in green cheese or put his feet down on something a lot more substantial. The Ranger project is primarily an unmanned photographic mission with lunar probes designed to impact on the moon, beginning with Ranger 1 and ending with Ranger 9. Basically, all the Ranger spacecraft are the same, with more sophisticated equipment added to the basic "bus" with each subsequent mission. Rangers 1 and 2 will test the engineering of spacecraft equipment and the system design, using a highly elliptical earth orbit. Ranger 1 is launched 8/15/61, but never reaches satellite orbit. Ranger 2 meets the same fate, but some scientific data is obtained. Rangers 3, 4, and

RANGER, SURVEYOR AND LUNAR ORBITER

Man gets his first close-up look at the moon

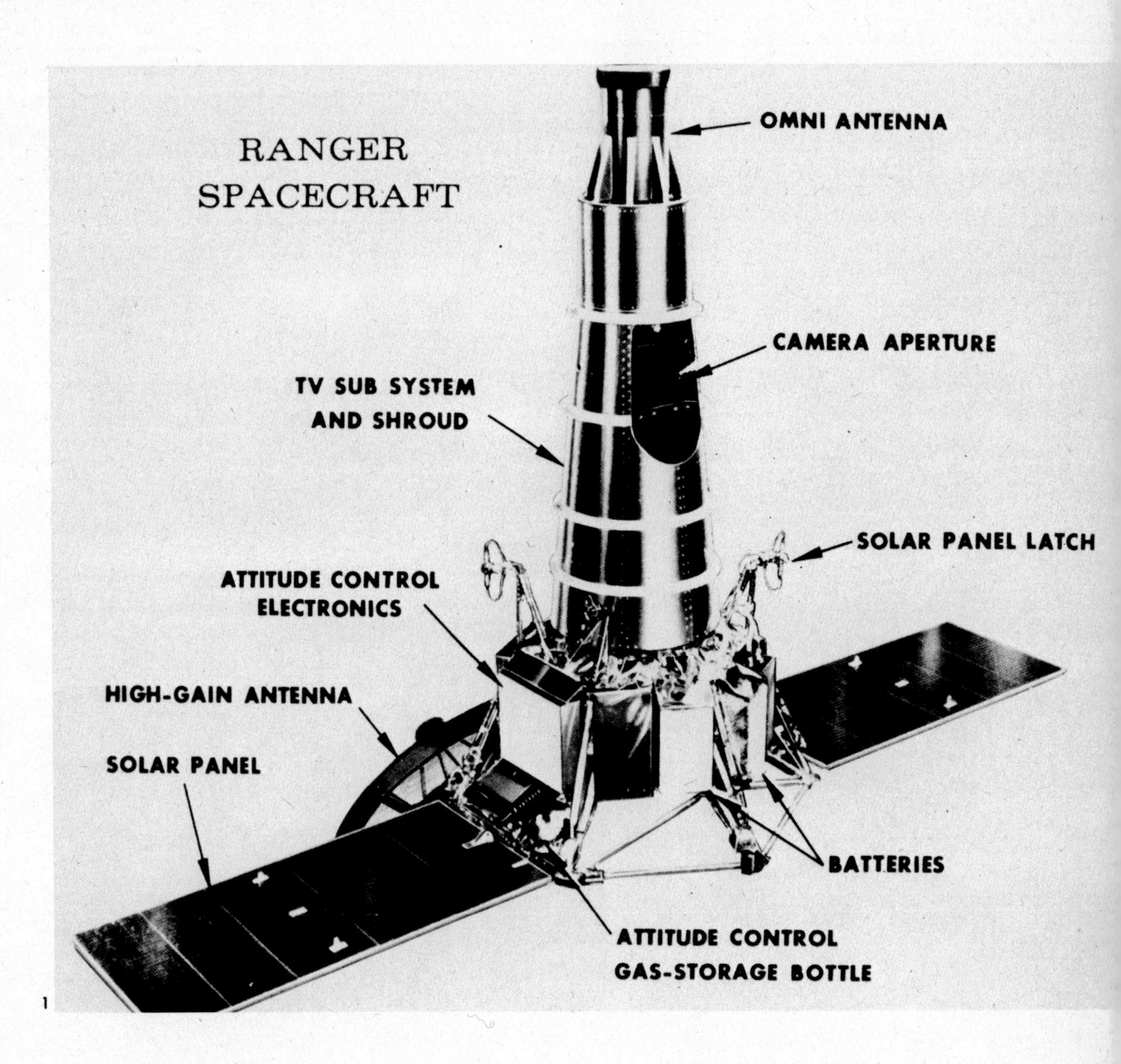

1

5 are launched in 1962, but Ranger 3 misses the moon by 22,862 miles; Ranger 4 impacts on the moon but suffers timer failure so that no photos are taken, and Ranger 5 misses the moon by 450 miles, although spacecraft transmitters continue sending for 11 days. Rangers 6 through 9 are planned for higher reliability. The new payload (**1, 2**) is to be a single scientific experiment. An interior view of Ranger 6 (**3**) shows the six TV cameras that will (hopefully) take and transmit high-resolution pictures at least ten times better than any taken from earth. While technicians assemble still another Ranger (**4**), the flight of Ranger 6 is marred when the high voltage power supply for the TV system accidentally turns on and destroys the entire television package, even though the payload impacts at the selected lunar site. The operation is a success, but the patient dies. Riding atop an Atlas/Agena launch vehicle, Ranger 7 (**5**) lifts off on July 7, 1964. The flight is perfect and Ranger 7, after transmitting more than 4300 pictures of the lunar surface, crashes into the moon, right on target. The last three pictures (**6**) were taken about ¼ second before impact; the point of impact is shown by a small white circle in each photo. The moment of impact (bottom, **6**) is shown by the strip of radio receiver noise on the right. Rangers 8 and 9 are also completely successful and between them transmitted nearly 13,000 pictures, closing the Ranger program in March of 1965. Never before have we seen such extreme close-up details of the surface.

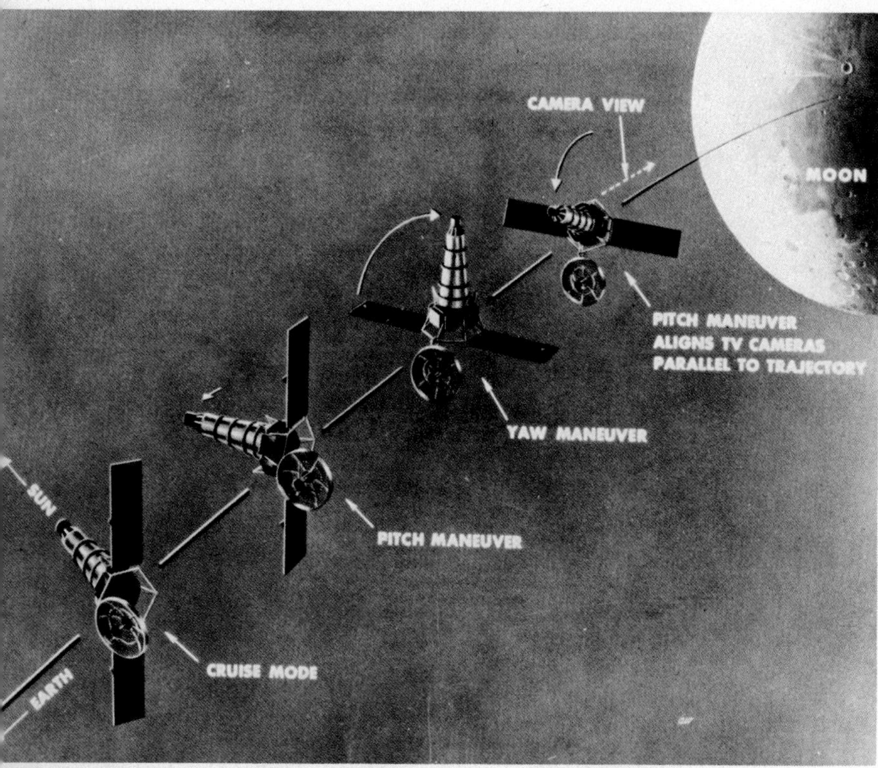

3

6

5

RANGER, SURVEYOR & LUNAR ORBITER

The purpose of the Ranger program was to photograph the surface of the moon and relay the pictures taken back to earth for study and analysis. The Surveyor program is to pick up where Ranger left off, and go several steps further. According to NASA, the program is slated to ". . . achieve soft landings on the moon by automated space-craft capable of transmitting scientific and engineering measurements from the lunar surface." In addition to developing and validating the technology for landing softly on the moon, and providing data on compatability of the Apollo manned lunar-landing spacecraft design, Surveyor findings will add to our scientific knowledge of the moon. The Surveyor spacecraft will all be basically the same, with additions and refinements with each subsequent flight. An artist's drawing (**1**) shows how Surveyor will look after soft-landing on the moon. The alpha-scattering instrument, which will test the properties of the soil, is in its stowed position. The surface sampler is extended to the surface,

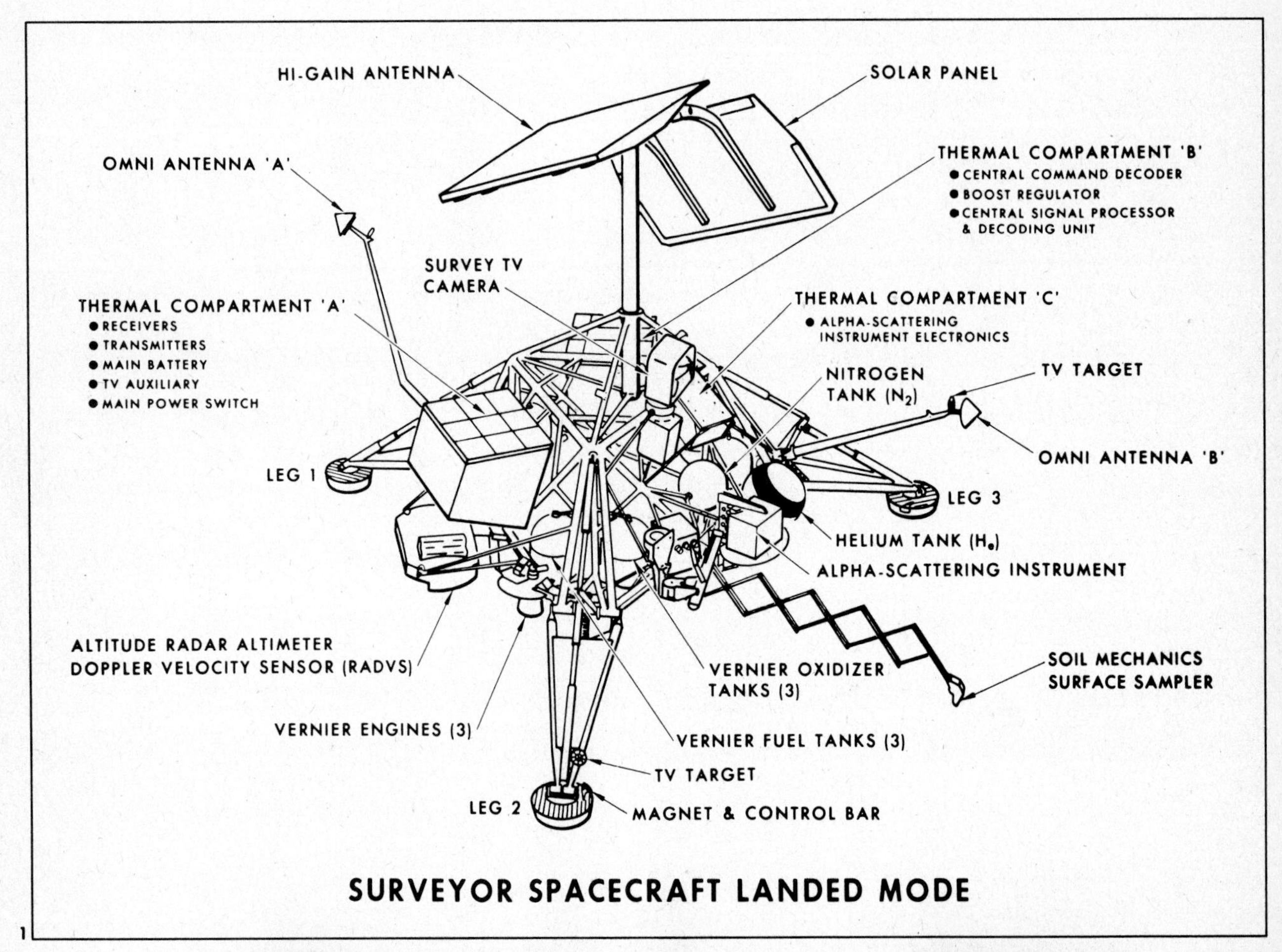

1

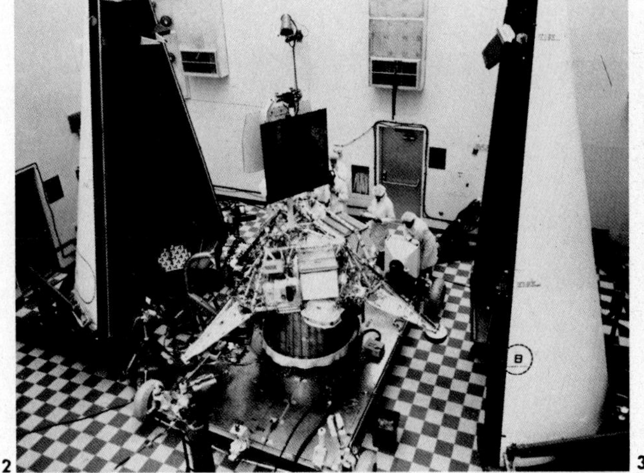
2

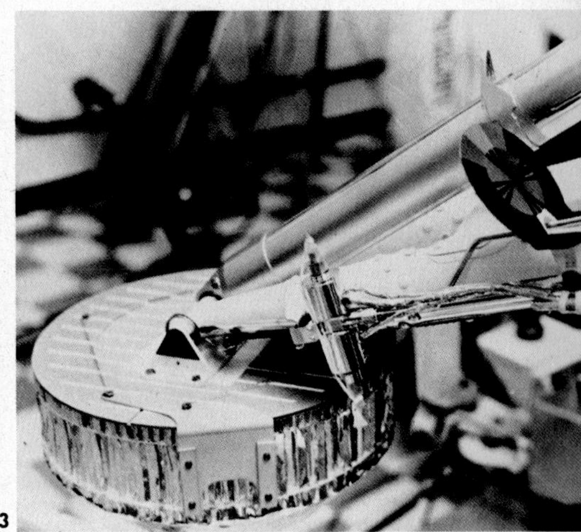
3

where it will expose subsurface material for television transmission. Technicians give Surveyor a final checkout (**2**) before it is enclosed in its protective nosecone shield. The hinged footpad of Surveyor 5 rests on the floor (**3**) before launch from Cape Kennedy. The wide tube is the landing leg; the narrow tube with the wires running along it is the shock absorber. The round disc positioned between the leg and shock absorber, looking like a phonograph record, is a TV target. The broken lines show how Surveyor 7 is scheduled to go to the moon (**4**). It is aimed for the moon's Central Bay at "X", and the drawing shows the position of Surveyor 7 in relation to the landing sites of Surveyors 1, 3, and 5. Surveyor 7 lifts off in a burst of flame on January 7, 1968 (**5**), headed for the crater Tycho; its soft-landing three days later is right on target and perfect. A mosaic of narrow-angle pictures taken by its television camera has been assembled (**6**) to show the surface sampler digging a trench. The camera, only four feet away from the lunar surface, shows Surveyor's other activities: the sensor head of the alpha-scattering instrument (upper left) was moved by the sampler onto the pile of soil excavated by the sampler scoop from two parallel trenches. Other results, such as trenches, bearing strength tests and disturbed rocks, can be seen elsewhere on the surface. With five successful flights out of seven, the Surveyor program has made significant discoveries. The lunar soil is dusty but firm, like fine-grained beach sand.

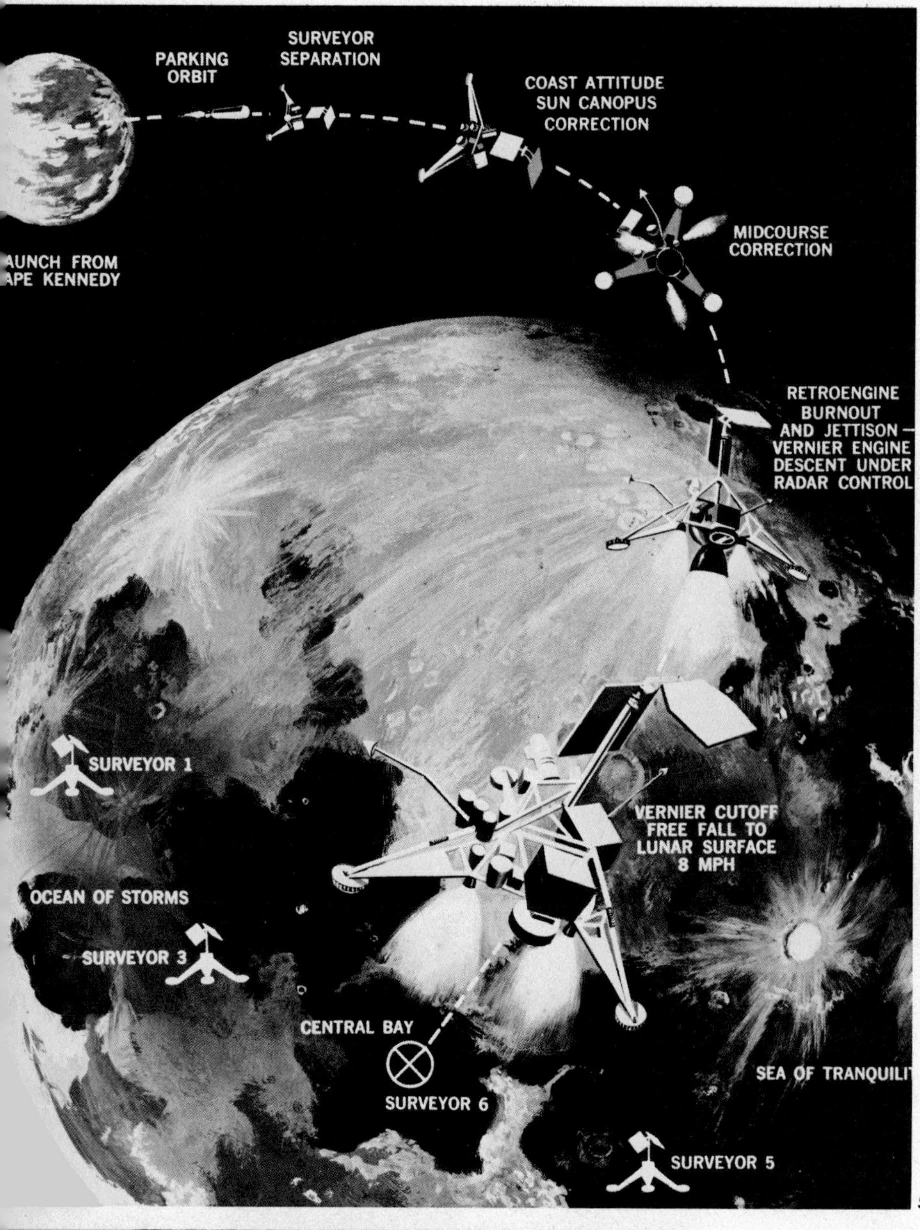

4

5

6

RANGER, SURVEYOR & LUNAR ORBITER

Project Lunar Orbiter begins in 1966 while the Ranger is still in operation. Five Lunar Orbiter spacecraft are to be launched into lunar orbit. The spacecraft will be almost identical, but each mission will be different in timing, trajectory, orbit, and target. The cameras in Lunar Orbiter will be used to develop a photographic map of the moon in order to select a suitable landing site for Apollo astronauts. Although the actual orbit may be different from the planned orbit, the Deep Space Network tracking and computation will establish the actual orbit by the time an Orbiter has made three passages around the moon. Then the Orbiter will be injected into an elliptical orbit that will bring it within 46 km. (27½ miles) of the moon's surface. Once Orbiter becomes a lunar satellite, its cameras will be able to map the lunar surface. In order to determine the height and slope of the moon's mountains and craters, the photography must take place when shadows are at a maximum, shortly after the lunar sunrise. Therefore, the spacecraft's or-

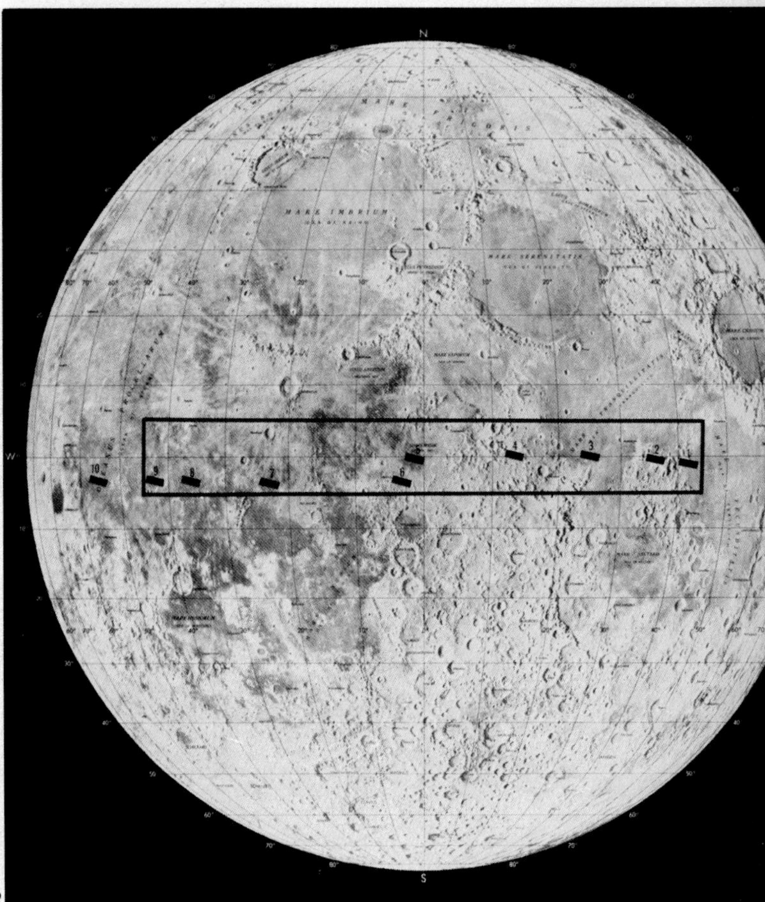

bit will vary only slightly with the sun's motion, as the low point (perilune) of the initial orbit is placed over the moon's sunrise zone. The moon's rotation on its axis will bring the target area under orbit and into the sunrise zone. An artist's drawing (**1**) depicts Lunar Orbiter 3 as it circles the moon. The ten small oblongs on the lunar surface (**2**) show the potential sites selected by NASA for photographing by Orbiter. The areas include examples of all the major types of lunar terrain for assessment of their suitability for spacecraft landings. A cutaway drawing (**3**) clearly shows all the parts of the spacecraft. By checking the drawing against the photo of the real thing (**4**), once can actually see what makes the Orbiter tick. At rest in the clean room at Kennedy Space Center, Lunar Orbiter (**5**) is mounted on a three axis test stand with its solar panels deployed. As with any other spacecraft, it will be enclosed in its protective cover and mounted atop an Atlas/Agena-D launch vehicle. Lunar Orbiter 3 lifts off from Cape Kennedy (**6**) on February 4, 1967. Considerable photo-mapping is accomplished during its successful flight. The program comes to a close with the photo-mapping of the far side of the moon by Lunar Orbiter 5, launched August 1, 1967. When all the information gathered has been evaluated, the site for the first manned lunar landing will be selected. Ranger's wide-angle photos, Orbiter's detailed photos, and Surveyor's tests and probings of the moon's barren soil provide abundant data for the crucial selection.

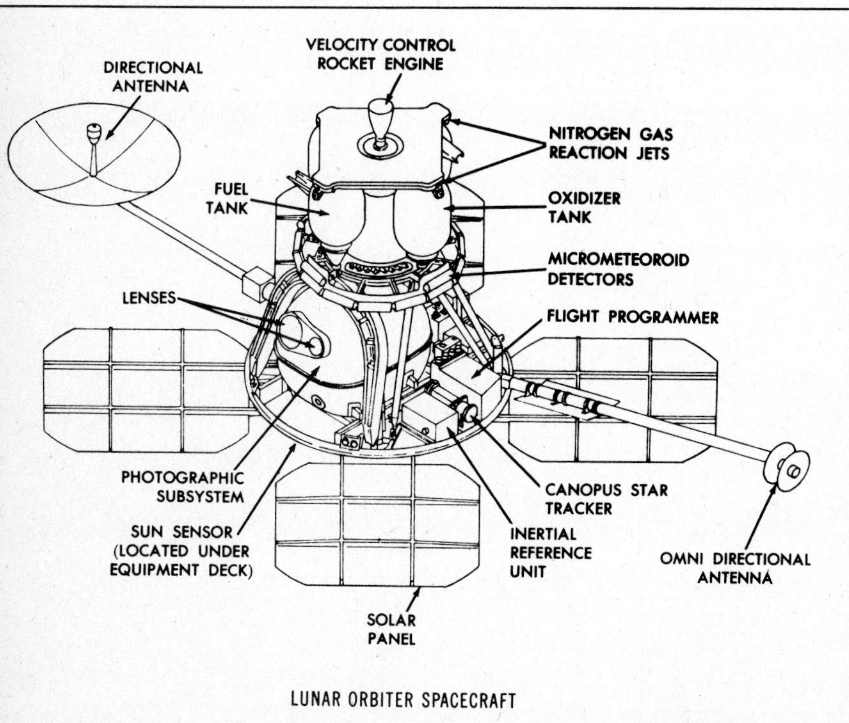

LUNAR ORBITER SPACECRAFT

6

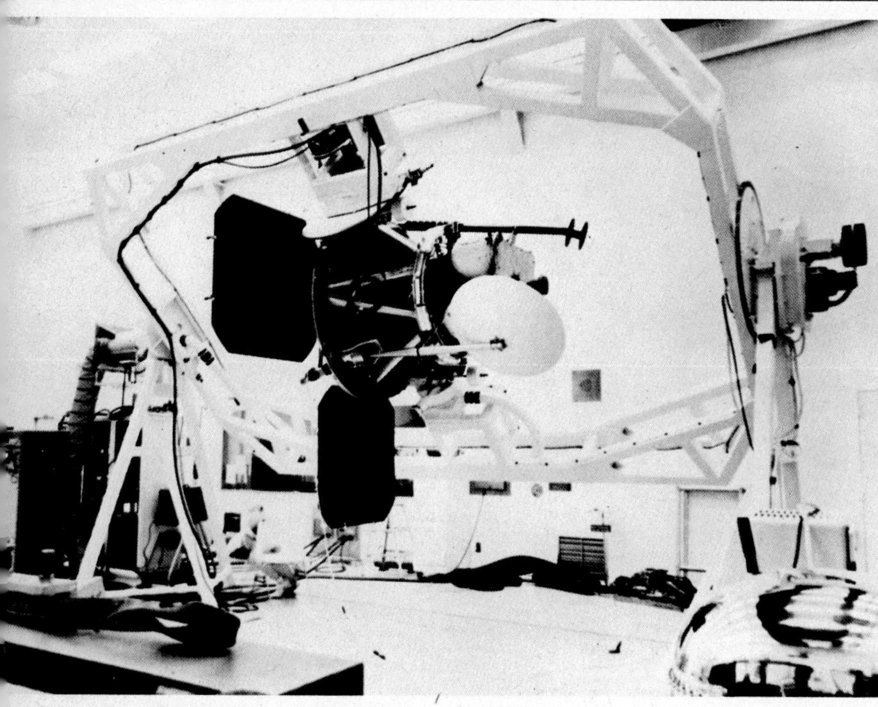

RANGER, SURVEYOR & LUNAR ORBITER

The Lunar Orbiter flight profile (**1**) reveals its complicated mission. Turning its wide angle lens on the southern half of the moon's hidden side, Orbiter 2 produced this remarkable view showing approximately 850,000 sq. miles of lunar surface (**2**). The picture was then transmitted to a tracking station at NASA's Deep Space Network. The small features detected at the top of the photo are about two-tenths of a mile across; the distance across the top is about 670 miles. The exposed camera film is developed in flight; a tiny beam of light scans the completed negative and converts it into electrical signals for radio transmission to earth. This "framelet" of primary Site 6a (**3**) was taken by Orbiter 2 with a 24-inch focal-length telephoto camera. The moon's far side (**4**) was captured in great detail at an altitude of about 900 miles. The sun is on the left, the terminator on the right. Crater Tsiolkovsky, first photographed by Russia's Luna 3 in 1959, is shown near the photo's center; it is named for the early Russian space theorist.

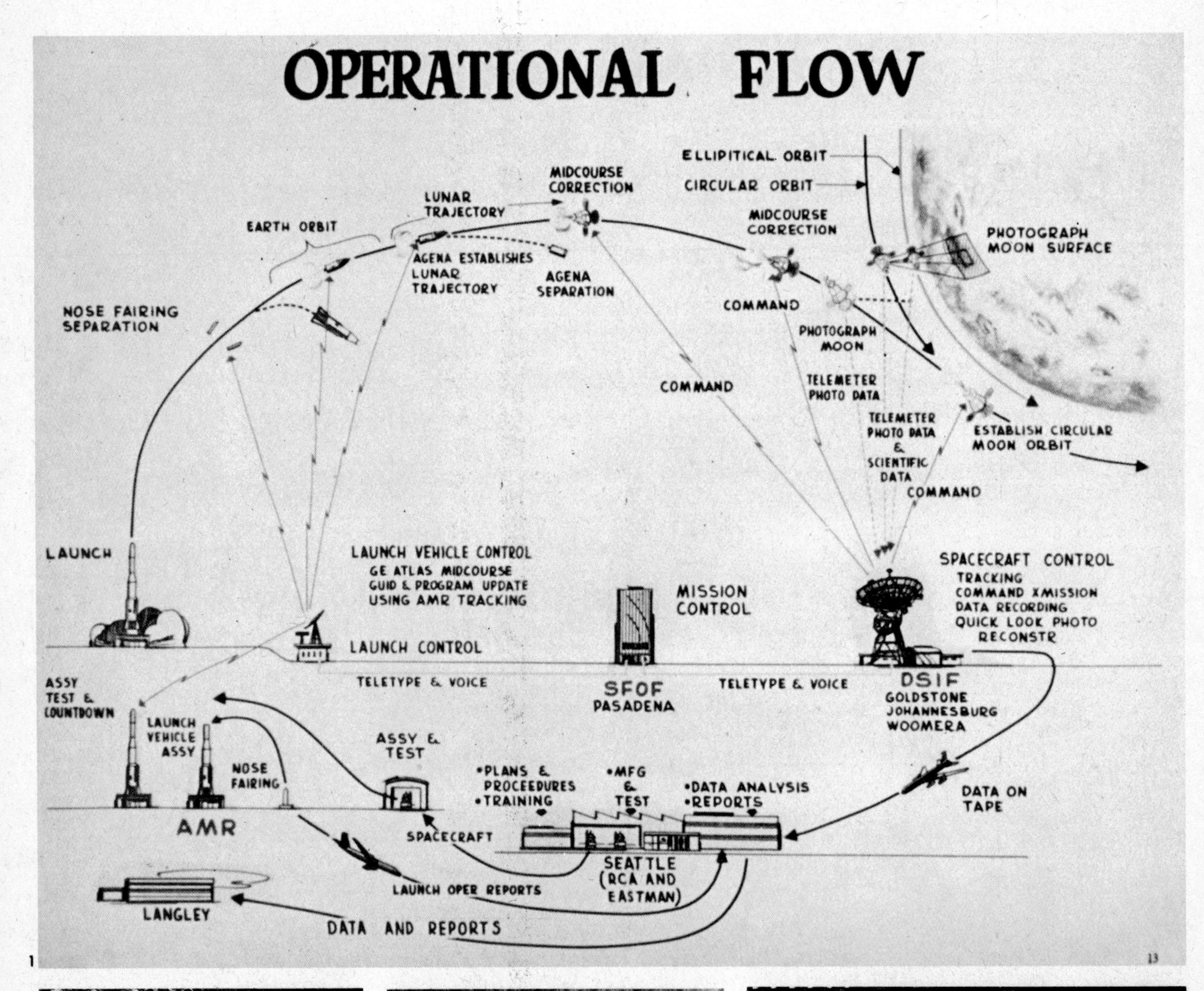

1

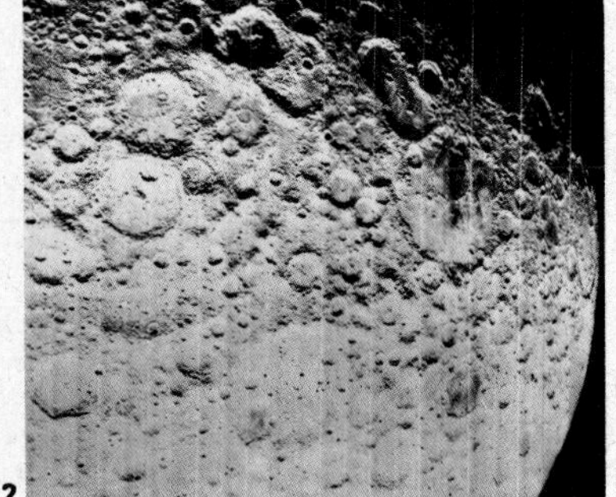

2

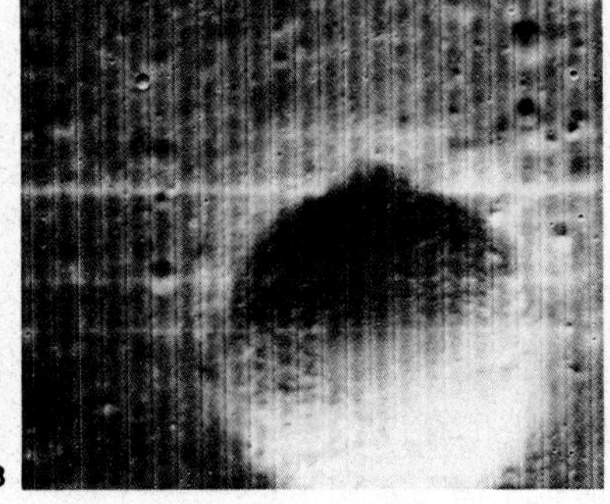

3

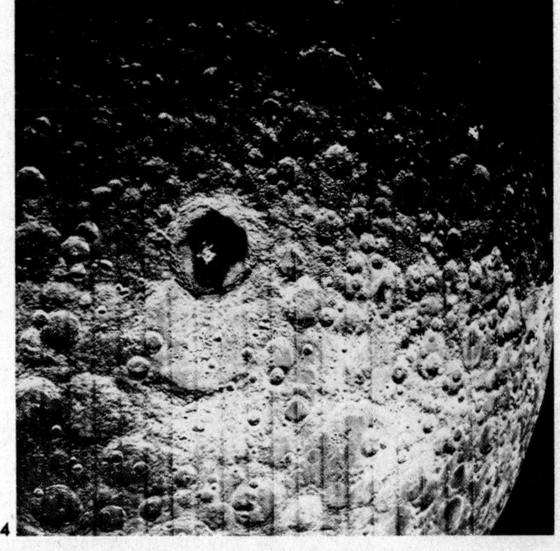

4